NATURAL-BORN CYBORGS

NATURAL-BORN CYBORGS

Minds, Technologies, and the Future of Human Intelligence

ANDY CLARK

OXFORD
UNIVERSITY PRESS
2003

OXFORD

UNIVERSITY PRESS

Oxford   New York
Auckland   Bangkok   Buenos Aires   Cape Town   Chennai
Dar es Salaam   Delhi   Hong Kong   Istanbul   Karachi   Kolkata
Kuala Lumpur   Madrid   Melbourne   Mexico City   Mumbai   Nairobi
São Paulo   Shanghai   Taipei   Tokyo   Toronto

Copyright © 2003 by Andrew J. Clark

Published by Oxford University Press, Inc.
198 Madison Avenue, New York, New York 10016

www.oup.com

Library of Congress Cataloging-in-Publication Data
Clark, Andy, 1957-
Natural-born cyborgs: Minds, technologies, and
the future of human intelligence / Andy Clark.
p. cm.
Includes bibliographical references and index.
ISBN 0-19-514866-5
1. Technology—Social aspects.
2. Neuroscience—Social aspects.
3. Artificial intelligence—Social aspects.
4. Human–computer interaction.
5. Cyborgs.
I. Title.
T14.5 .C58 2003
303.48'34—dc21
2002042521

1 3 5 7 9 8 6 4 2
Printed in the United States of America
on acid-free paper

For Mike Scaife, 1948–2001

Acknowledgments

This book owes large debts to many well-established ideas and research programs. All I have done is reshape these ideas, putting them into more direct contact with recent technological developments and with the ancient questions of who, what, and where we are. In constructing the foundations of this mosaic, I am most deeply indebted to the works of Daniel Dennett and Ed Hutchins. I also owe much to a brief but fruitful collaboration with David Chalmers (see our paper, "The Extended Mind" in *Analysis* 58, no. 1 [1998]: 7–19). In trying to see how specific new technologies fit in, I have been greatly helped by the works of Don Norman, Neil Gershenfeld, Kevin Kelly, Howard Rheingold, Yvonne Rogers, and Mike Scaife. Mike died, unexpectedly, while I was working on this book, and I respectfully dedicate it to his memory. Various other parts of the picture show the influence of Jerome Bruner, Richard Gregory, Donna Haraway, N. Katherine Hayles, David Kirsh, John Haugeland, Merlin Donald, Brian Arthur, Doug North, John Clippinger, Esther Thelen, and Linda Smith. Large but more subterranean influences include Merleau-Ponty, Heidegger, Lev Vygotsky, J. J. Gibson, Gregory Bateson, and Bruno Latour.

I was greatly inspired in the early days of this project by some interactions with N. Katherine Hayles, and with the organizers (especially Tom Foster, Louise Economides, and Laura Shackelford) of a round-table discussion that formed part of the Thinking Materiality workshop held at Indiana University, Bloomington, Indiana, in March 2000.

Works of fiction that had a special impact on me include pieces by Bernard Wolfe, Neil Stephenson, William Gibson, Bruce Sterling, Maureen McHugh, and Warren Ellis (*Limbo, Snow Crash, Neuromancer, Holy Fire, China Mountain Zhang,* and *Transmetropolitan,* respectively).

In keeping with my central theme, blame must also be shared by some of the key environments in which this work formed and developed. These include the School of Cognitive and Computing Sciences (and especially the Interact Lab) at the University of Sussex, UK; the Philosophy-Neuro-science-Psychology program, which I had the good fortune to direct for seven years at Washington University in St. Louis; the Santa Fe Institute; and most recently the Cognitive Science Program at Indiana University, Bloomington. Thanks too to Barbara Gorayska, Jacob Mey, Chrystopher Nehaniv, and the Cognitive Technology Society for involving me in their important work.

Some passages from the Introduction originally appeared in a short piece ("Natural-Born Cyborgs") electronically published by John Brockman as part of the *Edge*/Third Culture series. Thanks to John Brockman for permission to use this material here.

This book would not exist but for the support and encouragement of many people: my agents, John Brockman and Katinka Matson; Kirk Jensen of Oxford University Press; my wife and partner, Pepa Toribio; my mother and father, Christine and James Clark; Gill Banks; Miguel Toribio-Mateas, and all the close friends and family who have helped shape my thoughts and experiences over the years. My extra-large cat, Lolo, did a fair amount of shaping too.

Contents

NATURAL-BORN CYBORGS

The human skin is an artificial boundary: the world wanders into it, and the self wanders out of it, traffic is two-way and constant.

—Bernard Wolfe, *Limbo*

We're here to go.

—William S. Burroughs, *Dead City Radio*

Introduction

The Naked Cyborg

My body is an electronic virgin. I incorporate no silicon chips, no retinal or cochlear implants, no pacemaker. I don't even wear glasses (though I do wear clothes), but I am slowly becoming more and more a cyborg. So are you. Pretty soon, and still without the need for wires, surgery, or bodily alterations, we shall all be kin to the Terminator, to Eve 8, to Cable . . . just fill in your favorite fictional cyborg. Perhaps we already are. For we shall be cyborgs not in the merely superficial sense of combining flesh and wires but in the more profound sense of being human-technology symbionts: thinking and reasoning systems whose minds and selves are spread across biological brain and nonbiological circuitry. This book is the story of that transition and of its roots in some of the most basic and characteristic facts about human nature. For human beings, I want to convince you, are *natural-born* cyborgs.

This may sound like futuristic mumbo-jumbo, and I happily confess that I wrote the preceding paragraph with an eye to catching your attention, even if only by the somewhat dangerous route of courting your immediate disapproval! But I do believe that it is the plain and literal truth. I believe, to be clear, that it is above all a *SCIENTIFIC* truth, a reflection of some deep and important facts about (a whiff of paradox here?) our special, and distinctively *HUMAN,* nature. Certainly I don't think this tendency toward cognitive hybridization is a modern development. Rather, it is an aspect of our humanity, which is as basic and ancient as the use of speech and which has been extending its territory ever since. We see some of the "cognitive fossil trail" of the cyborg trait in the historical procession of potent cognitive technologies that begins with speech and counting, morphs first into written text and numerals, then into early printing (without moveable typefaces), on to the revolutions of moveable typefaces and the printing press, and most recently to the digital encodings that bring text, sound, and image into a uniform and widely transmissible format. Such technologies, once up and running in the various appliances and institutions that surround us, do far more than merely allow for the external storage and transmission of ideas. They constitute, I want to say, a cascade of "mindware upgrades": cognitive upheavals in which the effective architecture of the human mind is altered and transformed.

It was about five years ago that I first realized we were, at least in that specific sense, all cyborgs. At that time I was busy directing a new interdisciplinary program in philosophy, neuroscience, and psychology at Washington University in St. Louis. The realization wasn't painful; it was, oddly, reassuring. A lot of things now seemed to fall into place: why we humans are so deeply different from the other animals, while being, quite demonstrably, not so *very* different in our neural and bodily resources; why it was so hard to build a decent thinking robot; why the recent loss of my laptop had hit me like a sudden and somewhat vicious type of (hopefully transient) brain damage.

I'd encountered the idea that we were all cyborgs once or twice before, but usually in writings on gender or in postmodernist (or post postmodernist) studies of text. What struck me in July 1997 was that this kind of story was the literal and scientific truth. The human mind, if it is to be the physical organ of human reason, simply cannot be seen as bound and restricted by the biological skinbag. In fact, it has *never been* thus restricted and bound, at least not since the first meaningful words were uttered on some ancestral plain. But this ancient seepage has been gathering momentum with the advent of texts, PCs, coevolving software agents, and user-adaptive home and office devices. The mind is just less and less in the head.

If we do not always see this, or if the idea seems outlandish or absurd, that is because we are in the grip of a simple prejudice: the prejudice that whatever matters about *my* mind must depend solely on what goes on inside my own biological skin-bag, inside the ancient fortress of skin and skull. This fortress has been built to be breached; it is a structure whose virtue lies in part in its capacity to delicately gear its activities in order to collaborate with external, nonbiological sources of order to better solve the problems of survival and reproduction. It is because we are so prone to think that the mental action is all, or nearly all, on the inside, that we have developed sciences and images of the mind that are, in a fundamental sense, inadequate to their self-proclaimed target. So it is actually important to begin to see ourselves aright—it matters for our science, our morals, and our sense of self.

What, then, is the role of the biological brain, of those few pounds of squishy matter in your skull? The squishy matter is great at some things. It is expert at recognizing patterns, at perception, and at controlling physical actions, but it is not so well designed (as we'll see) for complex planning and long, intricate, derivations of consequences. It is, to put it bluntly, bad at logic and good at Frisbee. It is both our triumph and our burden, however, to have created a world so smart that it allows brains like ours to go where no animal brains have gone before. The story I want to tell is the story of that triumph, and of what it means for our understanding of ourselves: dumb thinkers in a smart world, or smart thinkers whose boundaries are simply not those of skin and skull?

The cyborg is a potent cultural icon of the late twentieth century. It conjures images of human-machine hybrids and the physical merging of flesh and electronic circuitry. My goal is to hijack that image and to reshape it, revealing it as a disguised vision of (oddly) our own biological nature. For what is special about human brains, and what best explains the distinctive features of human intelligence, is precisely their ability to enter into deep and complex relationships with nonbiological constructs, props, and aids. This ability, however, does not depend on physical wire-and-implant mergers, so much as on our openness to information-processing mergers. Such mergers may be consummated without the intrusion of silicon and wire into flesh and blood, as anyone who has felt himself thinking *via* the act of writing already knows. The familiar theme of "man the toolmaker" is thus taken one crucial step farther. Many of our tools are not just external props and aids, but they are deep and integral parts of the problem-solving systems we now identify as human intelligence. Such tools are best conceived as proper parts of the computational apparatus that constitutes our minds.

The point is best made by the series of extended concrete examples that I develop in this book. Consider, as a truly simplistic cameo, the process of using pen and paper to multiply large numbers.[1] The brain learns to make the most of its capacity for simple pattern completion ($4 \times 4 = 16$, $2 \times 7 = 14$, etc.) by acting in concert with pen and paper, storing the intermediate results outside the brain, then repeating the simple pattern completion process until the larger problem is solved. The brain thus dovetails its operation to the external symbolic resource. The reliable presence of such resources may become so deeply factored in that the biological brain alone is rendered unable to do the larger sums.

Some educationalists fear this consequence, but I shall celebrate it as the natural upshot of that which makes us such potent problem-solving systems. It is because our brains, more than those of any other animal on the planet, are primed to seek and consummate such intimate relations with nonbiological resources that we end up as bright and as capable of abstract thought as we are. It is because we are natural-born cyborgs, forever ready to merge our mental activities with the operations of pen, paper, and electronics, that we are able to understand the world as we do. There has been much written about our imminent "post-human" future, but if I am right, this is a dangerous and mistaken image. The very things that sometimes seem most post-human, the deepest and most profound of our potential biotechnological mergers, will reflect nothing so much as their thoroughly human source.

My cat Lolo is not a natural-born cyborg. This is so despite the fact that Lolo (unlike myself) actually does incorporate a small silicon chip. The chip is implanted below the skin of his neck and encodes a unique identifying bar code. The chip can be read by devices common in veterinarians' offices and animal shelters; it identifies me as Lolo's owner so we can be reunited if he is ever lost. The presence of this implanted device makes no difference to the shape of Lolo's mental life or the range of projects and endeavors he undertakes. Lolo currently shows no signs of cat-machine symbiosis, and for that I am grateful. By contrast it is our special character, as human beings, to be forever driven to create, co-opt, annex, and exploit nonbiological props and scaffoldings. We have been designed, by Mother Nature, to exploit deep neural plasticity in order to become one with our best and most reliable tools. Minds like ours were made for mergers. Tools-R-Us, and always have been.

New waves of user-sensitive technology will bring this age-old process to a climax, as our minds and identities become ever more deeply enmeshed in a nonbiological matrix of machines, tools, props, codes, and semi-intelligent daily objects. We humans have always been adept at dovetailing our minds and skills to the shape of our current tools and aids. But when those tools and aids start dovetailing back—when our technologies actively, automatically, and continually tailor themselves to us just as we do to them—then the line between tool and user becomes flimsy indeed. Such technologies will be less like tools and more like part of the mental apparatus of the person. They will remain tools in only the thin and ultimately paradoxical sense in which my own unconsciously operating neural structures (my hippocampus, my posterior parietal cortex) are tools. I do not really "use" my brain. There is no user quite so ephemeral. Rather, the operation of the brain makes me who and what I am. So too with these new waves of sensitive, interactive technologies. As our worlds become smarter and get to know us better and better, it becomes harder and harder to say where the world stops and the person begins.

Mind-expanding technologies come in a surprising variety of forms. They include the best of our old technologies: pen, paper, the pocket watch, the artist's sketchpad, and the old-time mathematician's slide rule. They include all the potent, portable machinery linking the user to an increasingly responsive world wide web. Very soon, they will include the gradual smartening-up and interconnection of the many everyday objects that populate our homes and offices.

However, this is not primarily a book about new technology. Rather, it is about us, about our sense of self, and about the nature of the human mind. It targets the complex, conflicted, and remarkably ill-understood relationship between biology, nature, culture, and technology. More a work of science-sensitive philosophy than a futurist manifesto, my goal is not to guess at what we might soon become but to better appreciate what we already are: *creatures whose minds are special precisely because they are tailor-made for multiple mergers and coalitions*.

All this adds important complexity to recent evolutionary psychological accounts that emphasize our ancestral environments.[2] We must take very seriously the profound effects of a plastic evolutionary overlay that yields a constantly moving target, an extended cognitive system whose constancy lies mainly in its continual openness to change. Even granting that the biological innovations that got this ball rolling may have consisted only in some small tweaks to an ancestral repertoire, the upshot of this subtle alteration is now a sudden, massive leap in the space of mind design. Our cognitive machinery is now intrinsically geared to self-transformation, artifact-based expansion, and a snowballing/bootstrapping process of computational and representational growth.

The line between biological self and technological world was, in fact, never very firm. Plasticity and multiplicity are our true constants, and new technologies merely dramatize our oldest puzzles (prosthetics and telepresence are just walking sticks and shouting, cyberspace is just one more place to be). Human intellectual history is, in large part, the tale of this fragile and always unstable frontier. The story I tell overlaps some familiar territory, touching on our skills as language-users, toolmakers, and tool-users. But it ends by challenging much of what we think we know about who we are, what we are, and even where we are. It ought to start, perhaps, somewhere on some dusty ancestral savanna, but join me instead on a contemporary city street, abuzz with the insistent trill of a hundred cell phones. . . .

Wired

Brighton main street, hub of a once-sleepy English seaside town lately transformed into a hi-tech haven and club-culture capital. This used to be my town, but it has changed. The shops tell a new story. I walk slowly, taking stock. I count one cell phone shop, one Starbucks, another cell phone shop, a hardware store, *another* cell phone shop, a clothes store, another coffee shop (this one offering full internet access), *yet another* cell phone shop . . .

The toll steadily mounts. Brighton, in my ten-year absence in the United States, has converted itself into a town that seems to sell nothing but coffee and cell phones. The center of town is now home to no fewer than fifty shops dedicated entirely to the selling of cell phones and their contracts. Then there are the various superstores that offer these phones alongside a variety of other goods. This is quite astonishing. For a relatively small town (around 250,000) this is surely a massive load. Yet business looks good and no wonder: everywhere I turn there are people with phone to ear, or punching in text messages using the fluent two-thumbed touch typing that is the badge of the younger users. Some, with fancier handsets, are using the phone to surf the web. This town is wired.

Not only is it wired. Half the people aren't entirely where they seem to be. I spent last Christmas in the company of a young professional whose phone was hardly ever out of his hands. He wasn't using the phone to speak but was constantly sending or receiving small text messages from his lover. Those thumbs were flying. Here was someone living a divided life: here in the room with us, but with a significant part of him strung out in almost constant, low-bandwidth (but apparently highly satisfying) contact with his distant friend.

The phone of the flying thumbs was a Nokia. Thanks in large part to Nokia (the firm, based in the Finnish town of the same name) the Finns emerged as early heavy-hitters in the European cell phone league. In 1999, 67 percent of the Finnish population owned and used cell phones compared to 28 percent in the United States. And these are not wimpy devices. Nokia is a pioneer of Wireless Application Protocol (WAP) technology, which supports fluent interfacing between the phone and the internet. Top of the line Finnish phones have for many years opened in the middle to reveal a small keyboard and screen supporting full fax, web, and e-mail capability. But it is not the potency of the technology so much as the pregnancy of the slang that really draws me to Finland. Finnish youngsters have dubbed the cell phone "kanny," which means extension of the hand.[3] The mobile is thus both something you use (as you use your hands to write) and something that is part of you. It is like a prosthetic limb over which you wield full and flexible control, and on which you eventually come to automatically rely in formulating and carrying out your daily goals and projects. Just as you take for granted your ability to use your vocal cords to speak to someone in the room beside you, you may take for granted your ability to use your thumbs-plus-mobile to send text to a distant lover. The phone really did seem to be part of the man, and the Finnish slang captures the mood.

I am surprised, but I shouldn't be. As a working cognitive scientist, the more I have learned about the brain and the mind, the more convinced I have become that the everyday notions of "minds" and "persons" pick out deeply plastic, open-ended systems—systems fully capable of including nonbiological props and aids as quite literally parts of themselves. No wonder the cell phone shops were full. These people were not just investing in new toys; they were buying *mindware upgrades*, electronic prostheses capable of extending and transforming their personal reach, thought, and vision.

Upgrades, as we all know, can be mixed blessings. Every new capacity brings new limits and demands. We may, for example, start to spread ourselves too thin, reconfiguring our work and social worlds in new and not necessarily better ways. Certainly, I felt more than a tad jealous of my friend's constant low-bandwidth info-dribble. It took some of him away from those he was physically beside. Later on, we'll take a closer look at some of these pros and cons in our cyborg future.

Brighton main street, then, is just one more sign of the times. As technology becomes portable, pervasive, reliable, flexible, and increasingly personalized, so our tools become more and more a part of who and what we are. With WAP-enhanced cell and access to our own personalized versions of the web in hand we see farther, organize better, know more. The temporary disability caused by a dead battery is unnerving. It seems we just aren't ourselves today. (The loss of my laptop, as I mentioned earlier, underlined this in a painfully personal way. I was left dazed, confused, and visibly enfeebled—the victim of the cyborg equivalent of a mild stroke.) So I, of all people, really *shouldn't* have been surprised. It is our natural proclivity for tool-based extension, and profound and repeated self-transformation, that explains how we humans can be *so very special* while at the same time being not so very different, biologically speaking, from the other animals with whom we share both the planet and most of our genes. What makes us distinctively human is our capacity to continually restructure and rebuild our own mental circuitry, courtesy of an empowering web of culture, education, technology, and artifacts. Minds like ours are complex, messy, contested, permeable, and constantly up for grabs. The neural difference that makes all this possible is probably not very large, but its effects are beyond measure.

Don't believe it yet? Or don't think it matters anyway? Both are fair and proper responses. I began deliberately with a technology—the cell phone—which is at once familiar yet insufficiently fluid and user-responsive to make (as yet) the strongest possible kind of case. And I have rehearsed none of the interlocking evidence (some philosophical, some psychological, some neuroscientific), which actually led me to embrace such a strong thesis in the first place.

Before the day is done, however, I hope to convince you at least of this: that the old puzzle, the mind-body problem, really involves a hidden third party. It is the mind-body-*scaffolding* problem. It is the problem of understanding how human thought and reason is born out of looping interactions between material brains, material bodies, and complex cultural and technological environments. We create these supportive environments, but they create us too. We exist, as the thinking things we are, only thanks to a baffling dance of brains, bodies, and cultural and technological scaffolding. Understanding this evolutionarily novel arrangement is crucial for our science, our morals, and our self-image both as persons and as a species.

CHAPTER 1

Cyborgs Unplugged

Rats in Space

The year is 1960. The pulse of space travel beats insistently within the temples of research and power, and the journal *Astronautics* publishes the paper that gave the term "cyborg" to the world.[1] The paper, titled "Cyborgs and Space," was based on a talk, "Drugs, Space and Cybernetics," presented that May to the Air Force School of Aviation Medicine in San Antonio, Texas. The authors were Manfred Clynes and Nathan Kline, both working for the Dynamic Simulation Laboratory (of which Kline was director) at Rockland State Hospital, New York. What Clynes and Kline proposed was simply a nice piece of lateral thinking. Instead of trying to provide artificial, earth-like environments for the human exploration of space, why not alter the humans so as to better cope with the new and alien demands? "Space travel," the authors wrote, "challenges mankind not only technologically, but also spiritually, in that it invites man to take an active part in his own biological evolution."[2] Why not, in short, reengineer the humans to fit the stars?

In 1960, of course, genetic engineering was just a gleam in science fiction's prescient eye. And these authors were not dreamers, just creative scientists engaged in matters of national (and international) importance. They were scientists, moreover, working and thinking on the crest of two major waves of innovative research: work in computing and electronic data-processing,[3] and work on cybernetics[4]—the science of control and communication in animals and machines. The way to go, they suggested, was to combine cybernetic and computational approaches so as to create man-machine hybrids, "artifact-organism systems" in which implanted electronic devices use bodily feedback signals to automatically regulate wakefulness, metabolism, respiration, heart rate, and other physiological functions in ways suited to some alien environment. The paper discussed specific artificial interventions that might enable a human body to bypass lung-based breathing, to compensate for the disorientations caused by weightlessness, to alter heart rate and temperature, reduce metabolism and required food intake, and so on.

It was Manfred Clynes who actually first suggested the term "cyborg." Clynes was at that time chief research scientist at Rockland State Hospital and an expert on the design and development of physiological measuring equipment. He had already received a prestigious Baker Award for work on the control of heart rate through breathing and would later invent the CAT computer, which is still used in many hospitals today. When Clynes coined the term "cyborg" to describe the kind of hybrid artifact-organism system they were envisaging, Kline remarked that it sounded "like a town in Denmark."[5] But the term was duly minted, and the languages of fact and fiction permanently altered. Here is the passage as it appeared in *Astronautics*:

> For the exogenously extended organizational complex . . . we propose the term "cyborg." The Cyborg deliberately incorporates exogenous components extending the self-regulating control function of the organism in order to adapt it to new environments.[6]

Thus, amid a welter of convoluted prose, was born the cyborg. The acronym "cyborg" stood for Cybernetic Organism or Cybernetically Controlled Organism; it was a term of art meant to capture both a notion of human-machine merging and the rather specific nature of the merging envisaged. Cyberneticists were especially interested in "self-regulating systems." These are systems in which the results of the system's own activity are "fed back" so as to increase, stop, start, or reduce the activity as conditions dictate. The flush/refill mechanism of a standard toilet is a homey example, as is the thermostat on the domestic furnace. The temperature drops, a circuit is activated, and the furnace comes to life. The temperature rises, a circuit is broken, and the furnace ceases to operate. Even more prosaically, the toilet is flushed, the ballcock drops, which causes the connected inlet valve to open. Water then flows in until the ballcock, riding on the rising tide, reaches a preset level and thus recloses the valve. Such systems are said to be homeostatically controlled because they respond automatically to deviations from a baseline (the norm, stasis, equilibrium) in ways that drag them back toward that original setting—the full cistern, the preset ambient temperature, and the like.

The human autonomic nervous system, it should be clear, is just such a self-regulating homeostatic engine. It works continuously, and without conscious effort on our part, in order to keep key physiological parameters within certain target zones. As effort increases and blood oxygenation falls, we breathe harder and our hearts beat faster, pumping more oxygen into the bloodstream. As effort decreases and blood oxygen levels rise, breathing and heart rate damp down, reducing the intake and uptake of oxygen.

With all this in mind, it is time to meet the first duly-accredited-and-labeled cyborg. Not a fictional monster, not even a human being fitted with a pacemaker (although they are cyborgs of this simple stripe too), but a white laboratory rat trailing an ungainly appendage—an implanted Rose osmotic pump. This rat (see fig 1.1) was introduced in the 1960 paper by Clynes and Kline as "one of the first cyborgs" and the snapshot, as Donna Haraway wonderfully commented "belongs in Man's family album."[7]

Sadly, the rat has no name, but the osmotic pump does. It is named after its inventor, Dr. Rose, who recently died after a very creative life devoted to

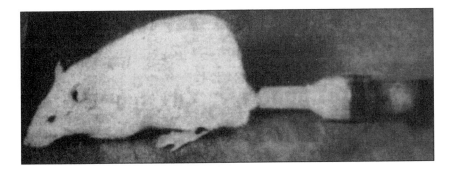

Fig. 1.1   An early (ca. 1955) classic cyborg: rat with implanted osmotic pump. The pump automatically injects chemicals into the rat to form a biotechnological control loop, which can be adapted to unusual conditions (for example, survival in space). By kind permission of Manfred Clynes.

the search for a cure for cancer. So let's respectfully borrow that, calling the whole rat-pump system Rose. Rose incorporates a pressure pump capsule capable of delivering injections at a controlled rate. The idea was to combine the implanted pump with an artificial control loop, creating in Rose a new layer of homeostasis. The new layer would operate like the biological ones without the need for any conscious attention or effort and might be geared to help Rose deal with specific extraterrestrial conditions. The authors speculate, for example, that the automatic, computerized control loop might monitor systolic blood pressure, compare it to some locally appropriate reference value, and administer adrenergic or vasodilatory drugs accordingly.

As cyborgs go, Rose, like the human being with the pacemaker, is probably a bit of a disappointment. To be sure, each incorporates an extra artificial layer of unconsciously regulated homeostatic control. But Rose remains pretty much a rat nonetheless, and one pacemaker doth not a Terminator make. Cyborgs, it seems, remain largely the stuff of science fiction, forty-some years of research and development notwithstanding.

Implant & Mergers

Or do they? Consider next the humble cochlear implant. Cochlear implants, which are already widely in use, electronically stimulate the auditory nerve. Such devices enable many profoundly deaf humans to hear again. However, they are currently limited by requiring the presence of a healthy, undegenerated auditory nerve. A Pasadena-based research group led by Douglas McCreery of Huntington Medical Research Institutes recently addressed this problem by building a new kind of implant (fig 1.2) that bypasses the auditory nerve and connects directly to the brain stem. Earlier versions of such devices have, in fact, been in use for a while, but performance was uninspiring. Uninspiring because these first wave brain stem implants used only an array of surface contacts—flat electrodes laid upon the surface of the brain stem near the ventral cochlear nucleus. The auditory discrimination of frequencies, however, is mediated by stacked layers of neural tissue within the nucleus. To utilize frequency information (to discriminate pitch) you need to feed information differentially into the various layers of this neural structure, where the stimulation of deeper layers results in the audi-

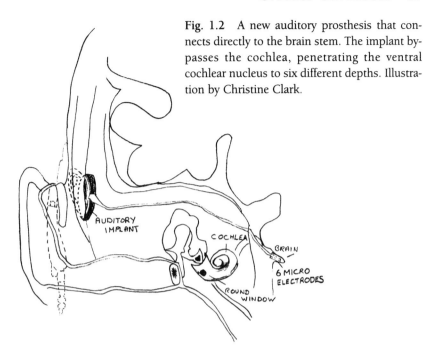

Fig. 1.2 A new auditory prosthesis that connects directly to the brain stem. The implant bypasses the cochlea, penetrating the ventral cochlear nucleus to six different depths. Illustration by Christine Clark.

tory perception of higher frequencies, and so on. The implant being pioneered by McCreery thus reaches deeper than those older, surface contact models, terminating in six iridium microelectrodes each of which penetrates the brain stem to a different depth. The overall system comprises an external speech processor with a receiver implanted under the scalp, directly wired to six different depths within the ventral cochlear nucleus. A Huntington Institute cat, according to neuroscientist and science writer Simon LeVay,[8] is already fitted with the new system and thus joins Rose in our Cyborg Hall of Fame.

The roll call would not be complete, however, without a certain maverick professor. Our next stop is thus the Department of Cybernetics at the University of Reading, in England. It is somewhat of a surprise to find, nowadays, a department of Cybernetics at all. They mostly died out in the early 1960s, to be replaced by departments of Computer Science, Cognitive Science, and Artificial Intelligence. But the real surprise is to find, within this Department of Cybernetics, a professor determined to turn himself into a good old-fashioned flesh-and-wires cyborg. The professor's name is Kevin Warwick, and in his own words:

I was born human. But this was an accident of fate—a condition merely of time and place. I believe it's something we have the power to change.[9]

Warwick began his personal transformation back in 1998, with the implantation of a fairly simple silicon chip, encased in a glass tube, under the skin and on top of the muscle in his left arm. This implant sent radio signals, via antennae placed strategically around the department, to a central computer that responded by opening doors as he approached, turning lights on and off, and so on. This was, of course, all pretty simple stuff and could have been much more easily achieved by the use of a simple device (a smart-badge or card) strapped to his belt or pinned to his lapel. The point of the experiment, however, was to test the capacity to send and receive signals via such an implant. It worked well, and Warwick reported that even in this simple case he quickly came to feel "like the implant was one with my body," to feel, indeed, that his biological body was just one aspect of a larger, more powerful and harmoniously operating system. He reported that it was hard to let go of the implant when the time came for its removal.

The real experiment took place on March 14, 2002, at 8:30 in the morning at the Radcliffe Infirmary, Oxford. There, Warwick received a new and more interesting implant. This consisted of a 100-spike array (see fig. 1.3).

Fig. 1.3   One-hundred-spike array, implanted into Professor Kevin Warwick, March 14, 2002 (shown against a small coin). By kind permission of Professor Warwick and of icube.co.uk.

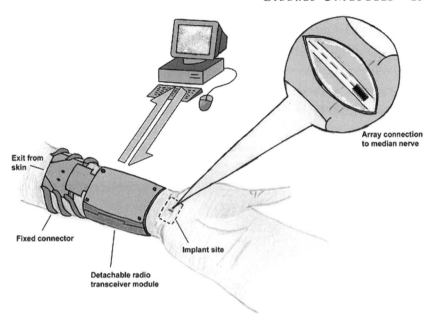

Array connection
to median nerve

Exit from
skin

Fixed connector

Implant site

Detachable radio
transceiver module

Fig. 1.4   Diagram of implant used by Professor Kevin Warwick. By kind permission of Professor Warwick.

Each of the 100 tips in the array makes direct contact with nerve fibers in the wrist and is linked to wires that tunnel up Professor Warwick's arm, emerging through a skin puncture where they are linked to a radio transmitter/receiver device (fig. 1.4). This allows the median nerve in the arm to be linked by radio contact to a computer. The nerve impulses running between brain and hand can thus be "wiretapped" and the signals copied to the computer. The process also runs in the other direction, allowing the computer to send signals (copies or transforms of the originals) to the implant, which in turn feeds them into the nerve bundles running between Warwick's hand and brain.

The choice of nerve bundles in the arm as interface point is doubtless a compromise. The surgical risks of direct neural interfacing are still quite high (the kind of brain stem implant described earlier, for example, is performed only on patients already requiring surgery to treat neurofibromatosis type 2). But the nerve bundles running through the arm do carry tremendous quantities of information to and from the brain, and they are implicated not just in reaching and grasping but also in the neurophysiology of pain, pleasure, and emotion. Warwick has embarked upon a staged sequence of experiments, the simplest of which is to record and identify the signals associated with specific willed hand motions. These signals can then be played back into his nervous system later on. Will his hand then move again? Will he feel as if he is willing it to move?

The experiment can be repeated with signals wiretapped during episodes of pain or pleasure. Warwick himself is fascinated by the transformative potential of the technology and wonders whether his nervous system, fed with computer-generated signals tracking some humanly undetectable quantity, such as infrared wavelengths, could learn to perceive them, yielding some sensation of seeing or feeling infrared (or ultraviolet, or x-rays, or ultrasound).[10]

Recalling the work on deep (cochlear nucleus penetrating) auditory repair, this kind of thing begins to seem distinctly feasible. Imagine, for example, being fitted with artificial sensors, tuned to detect frequencies currently beyond our reach, but sending signals deep into the developing ventral cochlear nucleus. Human neural plasticity, as we'll later see, may well prove great enough to allow our brains to learn to make use of such new kinds of sensory signal. Warwick is certainly enthusiastic. In his own words, "few people have even had their nervous systems linked to a computer, so the concept of sensing the world around us using more than our natural abilities is still science fiction. I'm hoping to change that."[11]

Finally, in a dramatic but perhaps inevitable twist, there is a plan (if all goes well) to subsequently have a matching but surface-level device connected to his wife, Irena. The signals accompanying actions, pains, and pleasures could then be copied between the two implants, allowing Irena's nervous system to be stimulated by Kevin's and vice versa. The couple also plans to try sending these signals over the internet, perhaps with one partner in London while the other is in New York.

None of this is really science fiction. Indeed, as Warwick is the first to point out, a great deal of closely related work has already been done. Scientists at the University of Tokyo have been able to control the movements of a live cockroach by hooking its motor neurons to a microprocessor; electronically mediated control of some muscular function (lost due to damage or disease) has been demonstrated in several laboratories; a paralyzed stroke patient, fitted with a neurally implanted transmitter, has been able to will a

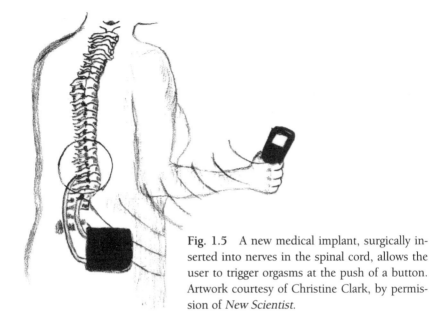

Fig. 1.5 A new medical implant, surgically inserted into nerves in the spinal cord, allows the user to trigger orgasms at the push of a button. Artwork courtesy of Christine Clark, by permission of *New Scientist.*

cursor to move across a computer screen; and rats with similar implants have learned to depress a reward-generating lever by just thinking about it.[12] There is even (fig. 1.5) a female orgasm-generating electronic implant (controlled by a hand-held remote) involving contacts surgically inserted into specific nerves in the spinal cord.[13] Without much doubt, direct bioelectronic signal exchanges, made possible by various kinds of implant technology, will soon open up new realms of human-computer interaction and facilitate new kinds of human-machine mergers. These technologies, for both moral and practical reasons, will probably remain, in the near future, largely in the province of restorative medicine or military applications (such as the McDonnell-Douglas Advanced Tactical Aircraft Program, which envisages a fighter plane pilot whose neural functions are linked directly into the on-board computer).[14]

Despite this, genuinely cyborg technology is all around us and is becoming more and more a part of us every day. To see why, we must reflect some more on what really matters even about the classic (wire-and-implant-dominated) cyborg technologies just reviewed. These classic cases all display direct (wire-based) animal-machine interfacing. Much of the thrill, or horror, depends on imagining all those wires, chips, and transmitters grafted onto pulsing organic matter. But what we should really *care about* is not the mere fact of deep implantation or flesh-to-wire grafting, but the complex and transformative nature of the animal-machine relationships that may or may not ensue. And once we see *that*, we open our eyes to a whole new world of cyborg technology.

Recall the case of the cochlear implants, and notice now the particular shape of this technological trajectory. It begins with simple cochlear implants connected to the auditory nerve—just one step up, really, from hearing aids and ear trumpets. Next, the auditory nerve is bypassed, and signals fed to contacts on the surface of the brain stem itself. Then, finally—classic cyborg heaven—microelectrodes actually penetrate the ventral cochlear nucleus itself at varying depths. Or consider Professor Warwick, whose first implant struck us as little more than a smart badge, worn inside the arm. My sense is that as the bioelectronic interface grows in complexity and moves inward, deeper into the brain and farther from the periphery of skin, bone, and sense organs, we become correlatively less and less resistant to the idea that we are trading in genuine cyborg technology.

But just why do we feel that depth *matters* here? It is, after all, pretty obvious that the physical depth of an implant, in and of itself, is insignificant. Recall my microchipped cat, Lolo. Lolo is, by all accounts, a disappointing cyborg. He incorporates a nonbiological component, conveniently placed within the relatively tamper-proof confines of the biological skin (and fur) bag. But he seems determinedly nontransformed by this uninvited bar coding. He is far from anyone's ideal of the cyborg cat. It would make no difference to *this* intuition, surely, were we to implant the bar code chip as deeply as we like—perhaps right in the center of his brain—humane technology and better bar code readers permitting. What we care about, then, is not depth of implanting per se. Instead, what matters to us is the nature and transformative potential of the bioelectronic coalition that results.

Still, the idea that truly profound biotechnological mergers must be consummated deep within the ancient skin-bag runs deep. It is the point source of the undeniable gut appeal of most classic cyborg technologies, whether real or imaginary. Think of adamantium skeletons, skull-guns, cochlear implants, retinal implants, human brains directly "jacked in" to the matrix of cyberspace—the list goes on and on.[15] The deeper within the biological skin-bag the bioelectronic interface lies, the happier we are, it seems, to admit that we confront a genuine instance of cyborg technology.

Intuitions, however, are strange and unstable things. Take the futuristic topless dancer depicted in Warren Ellis's wonderful and extraordinary *Transmetropolitan*.[16] The dancer (fig. 1.6) displays a fully functional three-inch-high bar code tattooed across both breasts. In some strange way, this merely superficially bar-coded dancer strikes me as a more unnerving, more genuinely cyborg image, than does the bar-coded cat. And this despite the fact that it is the latter who incorporates a genuine "within the skin-bag" implant. The reason for this reaction, I think, is that the image of the bar-coded topless dancer immediately conjures a powerful (and perhaps distressing) sense of a deeply transformed kind of human existence. The image foregrounds our potential status as trackable, commercially interesting sexual units, subject to repeated and perhaps uninvited electronic scrutiny. We resonate with terror, excitement, or both to the idea of ever-deeper neural and bodily implants in part *because* we sense some rough-and-ready (not foolproof, more of a rule-of-thumb) correlation between depth-of-interface and such transformative potential. The deep ventral cochlear nucleus penetrating implants

Fig. 1.6 Bar-coded dancer by Warren Ellis and Darick Robertson (detail from *Transmetropolitan 3*, Helix, DC Comics). By kind permission of Warren Ellis and Darick Robertson.

can, after all, upgrade the functionality of certain profoundly deaf patients in a much more dramatic, reliable, and effective fashion than its predecessors. What really counts is a kind of double whammy implicit in the classic cyborg image. First, we care about the potential of technology to become integrated so deeply and fluidly with our existing biological capacities and characteristics that we feel no boundary between ourselves and the nonbiological elements. Second, we care about the potential of such human-machine symbiosis to transform (for better or for worse) our lives, projects, and capacities.

A symbiotic relationship is an association of mutual benefit between different kinds of entities, such as fungi and trees. Such relationships can become so close and important that we tend to think of the result as a single entity. Lichen, for example, are really symbiotic associations between an alga and a fungus. It is often a vexed question how best to think of specific cases.[17] The case of cognitive systems is especially challenging since the requirement— (intuitive enough for noncognitive cases)—of physical cohesion within a clear inner/outer boundary seems less compelling when information flows (rather than the flow of blood or nutrients) are the key concern.

The traditional twin factors (of contained integration and profound transformation) come together perfectly in the classic cyborg image of the human body deeply penetrated by sensitively interfaced and capacity-enhancing electronics. But in the cognitive case, it is worth considering that what really matters might be just the *fluidity* of the human-machine integration and the resulting *transformation* of our capacities, projects, and lifestyles. It is then an empirical question whether the greatest usable bandwidth and potential lies with full implant technologies or with well-designed nonpenetrative modes of personal augmentation.[18] With regard to the critical features just mentioned, I believe that the most potent near-future technologies will be those that offer integration and transformation *without* implants or surgery: human-machine mergers that simply bypass, rather than penetrate, the old biological borders of skin and skull.

To see what I mean, let us return to the realms of the concrete and the everyday, scene-shifting to the flight deck of a modern aircraft. The modern flight deck, as the cognitive anthropologist Ed Hutchins has pointed out,[19] is designed as a single extended system made up of pilots, automated "fly-by-wire" computer control systems, and various high-level loops in which pilots monitor the computer while the computer monitors the pilots. The shape of these loops is still very much up for grabs. In the European Airbus,[20] the computer pretty much has the final say. The pilot moves the control stick, but the onboard electronics keep the flight deviations inside a preset envelope. The plane is not allowed, no matter what the pilots do with the control stick, to bank more than 67 degrees or to point the nose upward at more than 30 degrees. These computer-controlled limits are meant to keep the pilots' maneuvers from compromising the planes' structural integrity or initiating a stall. In the Boeing 747-400,[21] by contrast, the pilots still have the final say. In each case, however, under normal operating conditions, large amounts of responsibility are devolved to the computer-controlled autosystem. (The high-technology theorist and science writer Kevin Kelly nicely notes that human pilots are increasingly referred to, in professional training and talk, as "system managers.")[22]

Piloting a modern commercial airliner, it seems clear, is a task in which human brains and bodies act as elements in a larger, fluidly integrated, biotechnological problem-solving matrix. But still, you may say, this is state-of-the-art high technology. Perhaps there is a sense in which, at least while flying the plane, the pilots participate in a (temporary) kind of cyborg existence, allowing automated electronic circuits to, in the words of Clynes and Kline "provide an organizational system in which [certain] problems are taken care of automatically."[23] But most of us don't fly commercial airliners and are not even cyborgs for a day.

A Day in the Life

Or are we? Let's shift the scene again, this time to your morning commute to the office. At 7:30 A.M. you are awoken not by your native biorhythms but by your preset electronic alarm clock. By 8:30 A.M. you are on the road. It is a chilly day and you feel the car begin to skid on a patch of ice. Luckily, you have traction control and the Automatic Braking System (ABS). You simply hit the brakes, and the car takes care of most of the delicate work required. In fact, as we'll see in later chapters, the human brain is a past master at devolving responsibility in just this kind of way. You may consciously decide, for example, to reach for the wine glass. But all the delicate work of generating a sequence of muscle commands enabling precise and appropriate finger motions and gripping is then turned over to a dedicated, unconscious subsystem—a kind of on-board servomechanism not unlike those ABS brakes.

Arriving at your office, you resume work on the presentation you were preparing for today's meeting. First, you consult the fat file of papers marked "Designs for Living." It includes your own previous drafts, and a lot of work by others, all of it covered in marginalia. As you reinspect (for the umpteenth time) this nonbiological information store, your onboard wetware (i.e., your brain) kicks in with a few new ideas and comments, which you now add as supermarginalia on top of all the rest. Repressing a sigh you switch on your Mac G4, once again exposing your brain to stored material and coaxing it, once more, to respond with a few fragmentary hints and suggestions. Tired already—and it is only 10 A.M.—you fetch a strong espresso and go about your task with renewed vigor. You now position your biological brain to respond (piecemeal as ever) to a summarized list of key points culled from all those files. Satisfied with your work you address the meeting, presenting the final plan of action for which (you believe, card-carrying materialist that you are) your biological brain must be responsible. But in fact, and in the most natural way imaginable, your naked biological brain was no more responsible for that final plan of action than it was for avoiding the earlier skid. In each case, the real problem-solving engine was the larger, biotechnological matrix comprising (in the case at hand) the brain, the stacked papers, the previous marginalia, the electronic files, the operations of search provided by the Mac software, and so on, and so on. What the human brain is best at is learning to be a team player in a problem-solving field populated by an incredible variety of nonbiological props, scaffoldings, instruments, and resources. In this way ours are *essentially* the brains of natural-born cyborgs, ever-eager to dove-tail their activity to the increasingly complex technological envelopes in which they develop, mature, and operate.

What blinds us to our own increasingly cyborg nature is an ancient western prejudice—the tendency to think of the mind as so deeply special as to be distinct from the rest of the natural order. In these more materialist times, this prejudice does not always take the form of belief in soul or spirit. It emerges instead as the belief that there is something absolutely special about the cognitive machinery that happens to be housed within the primitive bioinsulation (nature's own duct-tape!) of skin and skull. What goes on in there is so special, we tend to think, that the only way to achieve a true human-machine merger is to consummate it with some brute-physical interfacing performed behind the bedroom doors of skin and skull.

However, there is nothing quite *that special* inside. The brain is, to be sure, an especially dense, complex, and important piece of cognitive machinery. It is in many ways special, but it is not special in the sense of providing a privileged arena such that certain operations must occur *inside* that arena, or in directly wired contact with it, on pain of not counting as part of our mental machinery at all. We are, in short, in the grip of a seductive but quite untenable illusion: the illusion that the mechanisms of mind and self can ultimately unfold only on some privileged stage marked out by the good old-fashioned skin-bag. My goal is to dispel this illusion, and to show how a complex matrix of brain, body, and technology can actually constitute the problem-solving machine that we should properly identify as *ourselves*. Seen in this light, the cell phones of the Introduction were not such a capricious choice of entry-point after all. None of us, to be sure, are yet likely to *think* of ourselves as born-again cyborgs, even if we invest in the most potent phone on the market and integrate its sweeping functionality deep into our lives. But the cell phone is, indeed, a prime, if entry-level, cyborg technology. It is a technology that may, indeed, turn out to mark a crucial transition point between the first (pen, paper, diagrams, and digital media dominated) and the second waves (marked by more personalized, online, dynamic biotechnological unions) of natural-born cyborgs.

Already, plans are afoot to use our cell phones to monitor vital signs (breathing and heart rate) by monitoring the subtle bounceback of the constantly emitted microwaves off of heart and lungs.[24] There is a simpler system, developed by the German company Biotronic, and already under trial in England, that uses an implanted sensor in the chest to monitor heart rate, communicating data to the patient's cell phone. The phone then automatically calls for help if heart troubles are detected. The list goes on.[25] The very designation of the mobile unit as primarily a phone is now in doubt, as more and more manufacturers see it instead as a multifunctional electronic bridge between the bearer and an invisible but potent universe of information, control, and response. At the time of writing, the Nokia 5510 combines phone, MP3 music player, FM radio, messaging machine, and game console, while Handspring's Trio incorporates a personal digital assistant. Sony Ericsson's T68i has a digital camera allowing the user to transmit or store color photos. Cell phones with integrated Bluetooth wireless technology (or similar) microchips will be able to exchange information automatically with nearby Bluetooth-enabled appliances. So enabled, a quick call home will allow the home computer to turn on or off lights, ovens, and other appliances.[26] In many parts of the world, the cell phone is already as integral to the daily routines of millions as the wristwatch—that little invention that let individuals take real control of their daily schedule, and without which many now feel lost and disoriented. And all this (in most cases) without a single incision or surgical implant. Perhaps, then, it is only our metabolically based obsession with our own skin-bags that has warped the popular image of the cyborg into that of a heavily electronically penetrated human body: a body dramatically transformed by prostheses, by neural implants, enhanced perceptual systems, and the full line of Terminator fashion accessories. The mistake—and it is a familiar one—was to assume that the most profound mergers and intimacies must always involve literal penetrations of the skin-bag.

Dovetailing

Nonpenetrative cyborg technology is all around us and is poised on the very brink of a revolution. By nonpenetrative cyborg technology I mean all the technological tricks and electronic aids that, as hinted earlier, are already transforming our lives, our projects, and our sense of our own capacities. What mattered *most*, even where dealing with real bioelectronic implants, was the potential for fluid integration and personal transformation. And while direct bioelectronic interfaces may contribute on both scores, there is another, equally compelling and less invasive, route to successful human-machine merger. It is a route upon which we as a society have already embarked, and there is no turning back. Its early manifestations are already part of our daily lives, and its ultimate transformative power is as great as that of its only serious technological predecessor—the printed word. It is closely related to what Mark Weiser, working at XeroxPARC back in 1988, first dubbed "ubiquitous computing" and what Apple's Alan Kay terms "Third Paradigm" computing.[27] More generally, it falls under the category of transparent technologies. Transparent technologies are those tools that become so well fitted to, and integrated with, our own lives and projects that they are (as Don Norman,[28] Weiser, and others insist) pretty much invisible-in-use. These tools or resources are usually no more the object of our conscious thought and reason than is the pen with which we write, the hand that holds it while writing, or the various neural subsystems that form the grip and guide the fingers. All three items, the pen, the hand, and the unconsciously operating neural mechanisms, are pretty much on a par. And it is this parity that ultimately blurs the line between the intelligent system and its best tools for thought and action. Just as drawing a firm line in this sand is unhelpful and misguided when dealing with our basic biological equipment so it is unhelpful and misguided when dealing with transparent technologies. For instance, do I merely *use* my hands, my hippocampus, my ventral cochlear nucleus, or are they part of the system— the "me"—that does the using?) There is no merger so intimate as that which is barely noticed.

Weiser's vision, ca. 1991, of ubiquitous computing was a vision in which our home and office environments become progressively more intelligent, courtesy of multiple modestly powerful but amazingly prolific intercommunicating electronic devices. These devices, many of which have since been produced and tested at XeroxPARC and elsewhere, range from tiny tabs to medium size pads to full size boards. The tabs themselves will give you the flavor. The idea of a tab is to "animate objects previously inert." Each book on your bookshelf, courtesy of its continuously active tab, would know where it is by communicating with sensors and transmitting devices in the building and office, what it is about, and maybe even who has recently been using it. Anyone needing the book can simply poll it for its current location and status (in use or not). It might even emit a small beep to help you find it on a crowded shelf! Such tiny, relatively dumb devices would communicate with larger, slightly less dumb ones, also scattered around the office and building. Even very familiar objects, such as the windows of a house, may gain new functionality, recording traces and trails of activity around the house. Spaces in the parking lot communicate their presence and location to the car-and-driver system via a small mirror display, and the coffee-maker in your office immediately knows when and where you have parked the car, and can prepare a hot beverage ready for your arrival.

The idea, then, is to embody and distribute the computation. Instead of focusing on making a richer and richer interface with an even more potent black box on the table, ubiquitous computing aims to make the interfaces multiple, natural, and so simple as to become rapidly invisible to the user.

The computer is thus drawn into the real world of daily objects and interactions where its activities and contributions become part of the unremarked backdrop upon which the biological brain and organism learn to depend.

This is a powerful and appealing vision. But what has it to do with the individual's status as a human-machine hybrid? Surely, I hear you saying, a smart world cannot a cyborg make. My answer: it depends just how smart the world is, and more importantly, how responsive it is, over time, to the activities and projects distinctive of an individual person. A smart world, which takes care of many of the functions that might otherwise occupy our conscious attention, is, in fact, already functioning very much like the cyborg of Clynes and Kline's original vision. The more closely the smart world becomes tailored to an individual's specific needs, habits, and preferences, the harder it will become to tell where that person stops and this tailor-made, co-evolving smart world begins. At the very limit, the smart world will function in such intimate harmony with the biological brain that drawing the line will serve no legal, moral, or social purpose. It would be as if someone tried to argue that the "real me" excludes all those nonconscious neural activities on which I so constantly depend relegating all this to a mere smart inner environment. The vision of the mind and self that remains following this exercise in cognitive amputation is thin indeed!

In what ways, then, might an electronically infested world come to exhibit the right kinds of boundary-blurring smarts? One kind of example, drawn from the realm of current commercial practice, is the use of increasingly responsive and sophisticated software agents. An example of a software agent would be a program that monitors your online reading and buying habits, and which searches out new items that fit your interests. More sophisticated software agents might monitor online auctions, bidding and selling on your behalf, or buy and sell your stocks and shares. Pattie Maes, who works on software agents at MIT media lab, describes them as

> software entities . . . that are typically long-lived, continuously running . . . and that can help you keep track of a certain task . . . so it's as if you were extending your brain or expanding your brain by having software entities out there that are almost part of you.

Reflect on the possibilities. Imagine that you begin using the web at the age of four. Dedicated software agents track and adapt to your emerging interests and random explorations. They then help direct your attention to new ideas, web pages, and products. Over the next seventy-some years you and your software agents are locked in a complex dance of co-evolutionary change and learning, each influencing, and being influenced by, the other. You come to expect and trust the input from the agents much as you expect and trust the input from your own unconscious brain—such as that sudden idea that it would be nice to go for a drive, or to buy a Beatles CD—ideas that seem to us to well up from nowhere but which clearly shape our lives and our sense of self. In such a case and in a very real sense, the software entities look less like part of your problem-solving environment than part of you. The intelligent system that now confronts the wider world is biological-you-plus-the-software-agents. These external bundles of code are contributing as do the various nonconscious cognitive mechanisms active in your own brain. They are constantly at work, contributing to your emerging psychological profile. You finally count as "using" the software agents only in the same attenuated and ultimately paradoxical way, for example, that you count as "using" your posterior parietal cortex.

The biological design innovations that make all this possible include the provision (in us) of an unusual degree of cortical plasticity and the (related) presence of an unusually extended period of development and learning (childhood).[29] These dual innovations (intensively studied by the new research program called "neural constructivism") enable the human brain, more than that of any other creature on the planet, to factor an open-ended set of biologically external operations and resources deep into its own basic modes of operation and functioning. It is the presence of this unusual plasticity that makes humans (but not dogs, cats, or elephants) *natural-born cyborgs*: beings primed by Mother Nature to annex wave upon wave of external elements and structures as part and parcel of their own extended minds.

This gradual interweaving of biological brains with nonbiological resources recapitulates, in a larger arena, the kind of sensitive co-development found within a single brain. A human brain, as we shall later see in more detail, comprises a variety of relatively distinct, but densely intercommunicating subsystems. Posterior parietal subsystems, to take an example mentioned earlier, operate unconsciously when we reach out to grasp an object, adjusting hand orientation and finger placement appropriately.[30] The conscious agent seldom bothers herself with these details: she simply decides to reach for the object, and does so, fluently and efficiently. The conscious parts of her brain learned long ago that they could simply count on the posterior parietal structures to kick in and fine-tune the reaching as needed. In just the same way, the conscious and unconscious parts of the brain learn to factor in the operation of various nonbiological tools and resources, creating an extended problem-solving matrix whose degree of fluid integration can sometimes rival that found within the brain itself.

Let's return, finally, to the place we started: the cyborg control of aspects of the autonomic nervous system. The functions of this system (the homeostatic control of heart rate, blood pressure, respiration, etc.) were the targets of Clynes and Kline in the original 1960 proposal. The cyborg, remember, was to be a human agent with some additional, machine-controlled, layers of automatic (homeostatic) functioning, allowing her to survive in alien or inhospitable environments. Such cyborgs, in the words of Clynes and Kline, would provide "an organizational system in which such robot-like problems were taken care of automatically, leaving man free to explore, to create, to think and to feel." Clynes and Kline were adamant that such off-loading of certain control functions to artificial devices would in no way change our nature as human beings. They would simply free the conscious mind to do other work.

This original vision, pioneering though it was, was also somewhat too narrow. It restricted the imagined cyborg innovations to those serving various kinds of bodily maintenance. There might be some kind of domino effect on our mental lives, freeing up conscious neural resources for better things, but that would be all. My claim, by contrast, is that various kinds of deep human-machine symbiosis really do expand and alter the shape of the psychological processes that make us who we are. The old technologies of pen and paper have deeply impacted the shape and form of biological reason in mature, literate brains. The presence of such technologies, and their modern and more responsive counterparts, does not merely act as a convenient wrap around for a fixed biological engine of reason. Nor does it merely free up neural resources. It provides instead an array of resources to which biological brains, as they learn and grow, will *dovetail* their own activities. The moral, for now, is simply that this process of fitting, tailoring, and factoring in leads to the creation of extended computational and mental organizations: reasoning and thinking systems distributed across brain, body, and world. And it is in the operation of these extended systems that much of our distinctive human intelligence inheres.

Such a point is not new, and has been well made by a variety of theorists working in many different traditions.[31] I believe, however, that the idea of human cognition as subsisting in a hybrid, extended architecture (one which includes aspects of the brain and of the cognitive technological envelope in which our brains develop and operate) remains vastly under-appreciated. We cannot understand what is special and distinctively powerful about human thought and reason by simply paying lip service to the importance of the web of surrounding structure. Instead, we need to understand in detail how brains like ours dovetail their problem-solving activities to these additional resources, and how the larger systems thus created operate, change, and evolve. In addition, we need to understand that the very ideas of minds and persons are not limited to the biological skin-bag, and that our sense of self, place, and potential are all malleable constructs ready to expand, change, or contract at surprisingly short notice.

Consider a little more closely the basic biological case. Our brains provide both some kind of substrate for conscious thought, and a vast panoply of thought and action guiding resources that operate quite unconsciously. You do not *will* the motions of each finger and joint muscle as you reach for the glass or as you return a tennis serve. You do not *decide* to stumble upon such-and-such a good idea for the business presentation. Instead, the idea just occurs to you, courtesy once again of all those unconsciously operating processes. But it would be absurd, unhelpful, and distortive to suggest that your true nature—the real "you," the real agent—is somehow defined only by the operation of the conscious resources, resources whose role may indeed be significantly less than we typically imagine. Rather, our nature as individual intelligent agents is determined by the full set of conscious and unconscious tendencies and capacities that together support the set of projects, interests, proclivities, and activities distinctive of a particular person. Just who we are, on that account, may be as much informed by the specific sociotechnological matrix in which the biological organism exists as by those various conscious and unconscious neural events that happen to occur inside the good old biological skin-bag.

Once we take all this on board, however, it becomes obvious that even the technologically mediated incorporation of additional layers of unconscious functionality must make a difference to our sense of who and what we are; as much of a difference, at times, as do some very large and important chunks of our own biological brain. Well-fitted transparent technologies have the potential to impact what we feel capable of doing, where we feel we are located, and what kinds of problems we find ourselves capable of solving. It is, of course, *also* possible to imagine bioelectronic manipulations, which quite directly affect the contents of conscious awareness. But direct accessibility to individual conscious awareness is not essential for a human-machine merger to have a profound impact on who and what we are. Indeed, as we saw, some of the most far-reaching near-future transformations may be rooted in mergers that make barely a ripple on the thin surface of our conscious awareness.

That this should be so is really no surprise. We already saw that what we cared about, even in the case of the classic cyborgs, was some combination of seamless integration and overall transformation. But the most seamless of all integrations, and the ones with the greatest potential to transform our lives and projects, are often precisely those that operate deep beneath the level of conscious awareness. New waves of almost invisible, user-sensitive, semi-intelligent, knowledge-based electronics and software are perfectly posed to merge seamlessly with individual biological brains. In so doing they will ultimately blur the boundary between the user and her knowledge-rich, responsive, unconsciously operating electronic environments. More and more parts of our worlds will come to share the moral and psychological status of parts of our brains. We are already primed by nature to dovetail our minds to our worlds. Once the world starts dovetailing back in earnest, the last few seams must burst, and we will stand revealed: cyborgs without surgery, symbionts without sutures.

CHAPTER 2

Technologies to Bond With

Heavy Metal

Los Alamos National Laboratory, New Mexico, occupies the high ground both physically and technologically. I am here, this hot and sunny day in May 1999, to deliver a talk on the interactions between mind and technology. Getting in is not easy. There have been security scares, and my permanent resident alien card is deemed insufficient proof of identity. After a flurry of panic, my secretary somehow manages to fax them a copy of my UK passport. At last—and just in time for the talk—I am issued with the inevitable plastic photo ID. Soon I find myself deep in the radiation-proof concrete bunkers that currently serve as home to my hosts, the Complex Systems Modeling Team.

Walking around this eerily silent, windowless, underground laboratory, I am struck by the stark contrast between old technology and new. The massive concrete bunkers and reinforced floors of these old buildings were designed both to resist nuclear attack and to support heavy, in-your-face technology: giant mainframes, immense monoliths of dials, lights, and levers. Yet today's action, in the Complex Systems Laboratory at least, usually requires little more than a few potent laptops and some fiber-optic links to massive databases. The heaviest piece of real, working machinery that I encounter is a somewhat sick old printer whose wheezing vibrations occasionally disturb the tomb-like silence.

The talk safely delivered, my hosts suggest a meal. Los Alamos' best restaurant turns out to be Japanese, an irony I decide not to pursue. But my political coyness proves unfounded. In fact, over the course of the meal it is decided that we will next visit one of Los Alamos' best, if lesser-known, attractions—the Black Hole.

The Black Hole is the shop-*cum*-soapbox of peace protester Edward Groshus.[1] To visit it is to step into a retro-technological Aladdin's Cave. Housed in a rambling, hangar-like complex on the edge of town, the Black Hole is a stunning repository of ex-National Laboratory equipment and scientific junk. The stuff was purchased (by the pound!) direct from the laboratory during its postwar sell-off period. The buyer was the same Ed Groshus, one-time national laboratory employee-turned-peacenik, anti-war campaigner, and retrotechnology entrepreneur. As I walk toward the installation, a billboard on the roadside catches my attention. It reads:

OMEGA CHURCH OF PEACE
BOMB UNWORSHIP CEREMONY
CRITICAL MASS EVERY SUNDAY

This is Groshus's doing. A man with an agenda, to be sure, but one nicely tempered by an enduring sense of fun. This shows, too, in his relationship with the goods in his store. Groshus despises the technologies of warfare, but he clearly sees the beauty as well as the absurdity of all that in-your-face technology. The Black Hole manages, incredibly, to be both shrine and protest. And it is a retro-techno addict's dream come true. It feels like a vast and ill-organized hardware store. The hardware here is not screws, nails, and duct tape so much as bank upon bank of imponderable valve electronics, heavyweight first-generation calculating machinery, fragments of complex control panels bristling with hundreds of tiny lights and switches, filters, fans, cathode-ray tubes, testers, probes, bomb casings, wires, screens, and dials. My personal favorite was a variety of gray, heavy, metal boxes (rather like office filing cabinets) with enormous single red buttons, labeled EMERGENCY, slap-bang in the middle, items seemingly straight out of Tom and Jerry, but in fact straight out of Uncle Sam, ca. 1960.

What we have here is an elephant's graveyard of Un-transparent, In-Your-Face Technology. Most of this stuff was not built to fade into the background of anyone's life or work. It made few efforts to configure itself to better suit the user. It was, in many ways, the strict antithesis of Weiser's vision of ubiquitous computing. Heavy, enormous, almost maximally resistant to easy human use, such technologies ran little risk of blurring the boundaries between machine and human, between biological user and technological tool. Naturally, I bought as much of it as I could possibly carry! I came away with a large vacuum tube with shining copper coil and a heavyweight electromagnet at the base. This was the so-called triggertron, once used to discharge a large bank of capacitors in order to implode trial atomic devices. I also succumbed to the siren call of two black boxes full of inscrutable, but wisely glowing, valve electronics. To complete the order I added a few substantial fragments of complex "Bat-Cave" control panels, featuring hundreds upon hundreds of tiny red and green lights and switches. To my eternal regret I could not carry the heavy box of metal with the big red emergency button in the middle. Walking all this through airport security at Albuquerque was surprisingly easy. "What's that, sir?" "It's an antique triggering device for an atomic weapon." "That's fine—come on through."

My own suspicious eroticization of retro-technology aside, the real point of this little reminiscence was just to begin to cement the contrast between two types of technology: "transparent technologies," and what might contrariwise be dubbed "opaque technologies." A transparent technology is a technology that is so well fitted to, and integrated with, our own lives, biological capacities, and projects as to become (as Mark Weiser and Donald Norman have both stressed) almost invisible in use.[2] An opaque technology, by contrast, is one that keeps tripping the user up, requires skills and capacities that do not come naturally to the biological organism, and thus remains the focus of attention even during routine problem-solving activity. Notice that "opaque," in this technical sense, does not mean "hard to understand" as much as "highly visible in use." I may not understand how my hippocampus works, but it is a great example of a transparent technology nonetheless. I may know exactly how my home PC works, but it is opaque (in this special sense) nonetheless, as it keeps crashing and getting in the way of what I want to do. In the case of such opaque technologies, we distinguish sharply and continuously between the user and the tool. The user's ongoing problem is to successfully deploy and control the tool. By contrast, once a technology is transparent, the conscious agent literally sees through the tool and directly confronts the real problem at hand. The accomplished writer, armed with pen and paper, usually pays no heed to the pen and paper tools while attempting to create an essay or a poem. They have become transparent equipment, tools whose use and functioning have become so deeply dovetailed to the biological system that there is a very real sense in which—while they are up and running—the problem-solving system just *is* the composite of the biological system and these nonbiological tools. The artist's sketch pad and the blind person's cane can come to function as transparent equipment, as may certain well-used and well-integrated items of higher technology, a teenager's cell phone perhaps. Sports equipment and musical instruments often fall into the same broad category.

Often, such integration and ease of use require training and practice. We are not born in command of the skills required. Nonetheless, some technologies may demand only skills that already suit our biological profiles, while others may demand skills that require extended training programs designed to bend the biological organism into shape. The processes by which a technology can become transparent thus include both natural fit (it requires only modest training to learn to use a hammer, for example) and the systematic effects of training. The line between opaque and transparent technologies is thus not always clear-cut; the user contributes as much as the tool. But there is a real and important sense in which some technologies are immediately better *candidates* for ultimate transparency than others. Most of the old heavyweight technology in the Black Hole remained eternally opaque, even to trained operators. While a very few devices are so well suited to the biological user that we either know at once how to use them, or quickly find out by an intuitive process of trial and error.

Transparent Tools

Donald Norman—cognitive scientist and contemporary guru of the age of "information appliances"—describes the Rubicon between opaque and transparent technologies in terms of a historical progression from "technology-centered" to "human-centered" products.[3] Human-centered products wear their functionality on their sleeve and exploit the natural strengths of human brains and bodies. These are the kinds of products where, Norman insists, the user almost never needs to open the manual. Yet vanishingly few of our high technology, information-based products are like that today. Sadly, a thousand cases of highly opaque, run-daily-to-the-manual products spring all-too-readily to mind. Examples range from VCRs to photocopiers to personal organizers and laptops.

The trouble with technology-centered products is that, as Norman's label suggests, they answer only to the need to do things (often, many *different* things) that previous products didn't do, or that they didn't do to the same degree. What they *don't* answer to is the need to enable those things to be done fluently, reliably, and with a minimum of learning and effort on the part of the user. And the reason, as Norman notes, is simple enough. At first, creating a product that can DO THE JOB is hard enough, let alone aiming for products nicely fitted to brains like ours. As time goes by, however, the vendors must seek to extend their market beyond the gung ho early adopters and technophiles. They will need to sell to the average user who simply wants a cheap, reliable, and easy-to-use tool. The technological product then comes under cultural-evolutionary pressure to increase its fitness by better conforming to the physical and cognitive strengths and weaknesses of biological bodies and brains. In quasi-evolutionary terms, the product is now poised to enter into a kind of symbiotic relationship with its biological users. It requires widespread adoption by users if its technological lineage is to continue, and one good way to achieve this is to provide clear benefits at low cognitive and economic costs.

There are, of course, many rather less appealing routes to technological (and biological) survival: products, for example, which survive simply because they do the job, however opaquely, or products that depend on the incompatibility of alternatives with some popular platform or protocol. There is also a large gray area (discussed further in chapter 7) in which technologies actively create the very needs (e.g., "more memory") that they then rush to fulfill. Our immediate task, however, is to get a more concrete sense of some of the complex ways in which technologies simultaneously shape and adapt to the cognitive profiles of biological users. With that in mind, let's look briefly at a familiar item, one that long-ago passed from the realm of opaque technology into that of transparent symbiotic partner— the humble wristwatch.[4]

We humans didn't always keep precise, objectively measured time. Before the dawn of the city, the factory, and the organized religious order, human beings used natural cycles to prompt daily activities. The sun rises and farming begins, interrupted only by a brief break when the sun is high in the sky. Darkness signals food and sleep. Today, a great many humans are not like this. We work all hours. We plan to meet friends for coffee at 11:45 A.M. We make a date for supper at 10:00 P.M. and a film at midnight, and so on. The transition from a natural-time society to our present arrangements for work and play was mediated by a long thread of technological evolution: a thread that leads from heavy, fixed, unreliable sundials and water clocks, through the development of early oscillating-element-based timekeeping, right up to cheap, accurate, personal quartz crystal wristwatches. But the technological story, though fascinating, pales beside the human-centered story. In a mere five hundred years, the opaque, unreliable, fixed-location tower clocks of the Middle Ages gave way to the reliable, cheap, personal timekeepers that we now take so much for granted. Along the way our relationship to time itself was irrevocably changed and transformed.

Once the average city worker was awakened by the call of the night watch, a living person whose task was to patrol the streets shouting the time. A little later the tolling of a bell, either owned by the town or perhaps by a specific employer, woke the townspeople. These measures instilled a degree of what David Landes nicely calls "time obedience." But with the availability of *personal* timepieces, in the form of chamber clocks or (ultimately) wristwatches, came the possibility of something new and different—"time discipline." The presence of easily accessible, fairly accurate, and consistently available time-telling resources enabled the individual to factor time constantly and accurately into the very heart of her endeavors and aspirations. This made possible ways of thought, and cultural practices and institutions, which were otherwise precluded by our basic biological nature. Landes makes the point well:

> The public clock could be used to open markets and close them, to signal the start of work and its end, to move people around, but it was a limited guide to self-imposed programs. Its dial was not always in view; its bells not always within hearing. Even when heard, hourly bells are at best intermit-

tent reminders. They signal moments. A chamber clock or watch is something very different: an ever-visible, ever-audible companion and monitor . . . a measure of time used, time spent, time wasted, time lost. As such it was prod and key to personal achievement and productivity.[5]

Notice that what counts here is not always *consciously knowing* the time. None of us, I suppose, looks constantly at his or her watch! Rather, the crucial factor is the constant and easy availability of the time, *should we desire to know it*. Therefore, a prime characteristic of transparent technologies is their *poise* for easy use and deployment as and when required. Daily, unreflective usage bears this out. As you walk down the street, you are accosted by the familiar cry of the temporarily watchless. "Excuse me, sir, do you happen to know the time?" Asked this question on a busy street, most of us will unhesitatingly reply, *even before consulting our wristwatches*, that yes, we surely do.[6] Grasping the request hidden in the formulaic question, many of us will also, and without further request, share our knowledge with the time-challenged supplicant. As we do so, we may find ourselves producing one of the characteristic body motions of the modern world. In the suited male or female, this takes the form of a controlled, punch-like extension of the arm, a clockwise half-rotation of the emerging wrist, and a slight lowering of the gaze. This knowledge-retrieval tropism serves, of course, a single practical function—it permits you to focus your gaze briefly upon the face, dial, or display of your watch, that humble example of cyborg technology.[7]

Now compare a superficially similar case. Your houseguest has encountered a word he does not know. To be concrete, let the word be "clepsydra." At some appropriate conversational juncture, the question is raised: "Good host, do you know what the word 'clepsydra' means?" Perhaps you are like me. I only learned this word a few days before writing this paragraph; until then, it wasn't part of my working vocabulary at all. But perhaps, like me, you keep a medium-size version of the *Oxford English Dictionary* somewhere in your house. So you know you have the wherewithal to resolve the matter. But what do you say? You surely *won't* say "Yes, I know what that word means" and only then proceed to consult the dictionary. Yet this is precisely what usually happens when we are asked the time!

An easy dismissal of this discrepancy is, of course, to simply lay everything at the accommodating feet of convention. When we answer that we know the time, all we mean is that we have the information readily at hand. And to be sure, several cultural variants of the request exist. My wife, a native Spanish speaker, might ask me *"Tienes hora?"* literally, "Have you got the time?" with the emphasis on possession rather than knowledge. All this notwithstanding, I think the ease with which we accept talk of the watch-bearer as one who actually knows—rather than one who can easily find out—the time is suggestive. For the line between that which is *easily and readily accessible* and that which should be counted as *part of the knowledge base* of an active intelligent system is slim and unstable indeed. It is so slim and unstable, in fact, that it sometimes makes both social and scientific sense to think of your individual knowledge as quite simply whatever body of information and understanding is at your fingertips; whatever body of information and understanding is right there, cheaply and easily available, as and when needed.[8] According to one diagnosis, then, you are telling the literal truth when you answer "yes" to the innocent-sounding question "Do you know the time?" For you *do* know the time. It is just that the "you" that knows the time is no longer the bare biological organism but the hybrid biotechnological system that now includes the wristwatch as a proper part.

To make this just a little more palatable, consider the parallel case of biological memory. Suppose I ask you whether you know the year of the first walk on the moon. You might answer "Yes, 1969." In answering "yes," you do not mean to imply that this date was present to your conscious awareness all along. You do not walk around all day mentally rehearsing "1969," "1969," "1969." Rather, your "yes" signifies that the information was indeed there, poised for easy access and retrieval from your biological memory. The informational poise of the wristwatch (and, as we'll later see, of the visual scene in front of your own eyes) may sometimes be relevantly similar. Perhaps, then, you may be properly said to know the time even before you actually look at your watch—just as you can be said to know the date of the moon landing even before actually retrieving it from your biological memory.

If this way of looking at things still strikes you as outlandish, you are in good company. Most people find such a diagnosis strange, unnecessary and (thus) unconvincing. But this reaction is unprincipled. It rests not upon any deep fact about the nature of knowledge or the preset bounds of persons but on a simple prejudice: the contemporary version, as it happens, of the old and discredited idea of the mind as a special kind of spirit-stuff. The idea of "mind as spirit-stuff" is no longer scientifically respectable. Instead, mind is seen as the working of a purely physical device. In identifying that physical device solely with the biological brain, we again make a leap of faith, depicting the biological brain itself as the sole and essentially *insulated* engine of mind and reason. This conception is the old idea of special spirit-stuff in modern dress. A thoroughgoing physicalism should allow mind to determine—by its characteristic actions, capacities, and effects—its own place and location in the natural order. We should not, at any rate, simply assume that it is correct to identify and locate the individual *thinking system* by reference to the merely *metabolic* frontiers of skin and skull.

We can, in any event, take away two somewhat less contentious lessons from our discussion of modern timekeeping. The first is that transparent (nonopaque, human-centered) technology is by no means a new invention. It is with us already in a wide variety of old technologies, including pen, paper, books, watches, written words, numerical notations, and the multitude of almost-invisible props and aids that scaffold and empower our daily thought and action.[9] The second is that the passage to transparency often involves a delicate and temporally extended process of co-evolution. Certainly, the technology must change in order to become increasingly easy to use, access, and purchase; but this is only half the story because at the same time, elements of culture, education, and society must change also. In the case at hand, people had to learn to *value* time discipline as opposed to mere time obedience, and this transition itself, Landes tells us, took over a hundred years to fully accomplish.

Smart Worlds

What happened with timekeeping is now happening with the flow of information itself. Mark Weiser's vision of ubiquitous computing is finding concrete expression in attempts to design and market what Norman calls "information appliances."[10] We have met this phrase once or twice already,

and it is time to try to pin it down. Information appliances are character-
ized by three central features:

1. An information appliance is geared to support a specific activity, and to do
   so via the storage, reception, processing, and transmission of information.

2. Information appliances form an intercommunicating web. They can "talk"
   to each other.

3. Information appliances are transparent technologies, designed to be easy
   to use, and to fade into the background. They are *poised to be taken for
   granted*.[11]

Weiser's vision of the home and workplace as filled with small, intercommunicating, unobtrusive intelligent devices was a vision of a world of
such appliances, but Norman offers several rather more restricted, less futuristic, examples. He imagines the use of inexpensive, tiny cameras to
beam information (the shape of the coffee table, the color of the sweatshirt)
directly to family and friends while shopping. This is a natural extension of
the current use of cell phone technology. (In fact, as I write the second
draft of this text, I note that several cell phone companies offer "picture-
messaging" with attendant ability for personal digital input.) Norman goes
on to imagine houses with permanent wall-mounted weather displays, constantly showing the local forecast and conditions, to imagine (echoing
Weiser) applications embedded in walls and furniture, and supermarkets
where you simply wheel the laden shopping cart through a sensor, which
scans each item and debits your bank account accordingly. He imagines
devices embedded in our clothes, eyeglasses capable of comparing a currently presented face to a database and retrieving name and details (again,
I lately discover that such glasses now exist and are being marketed as aids
for mild Alzheimer's sufferers). And he imagines—the inevitable final step—
similar devices implanted in our own bodies, monitoring the world, communicating with other such devices, and enabling us to manage, recognize,
store, and compare information quite effortlessly as we go about our daily
business.

Such is Norman's vision: a vision of a world in which "information is
more available to all of us, no matter where we are, whenever we need it."[12]
Such technologies, to support the kind of profound integration into hu-

man life here envisaged, need to be just about maximally nonopaque. They should contribute nothing to the complexity of the tasks they support: "the complexity of the appliance is that of the task, not the tool."[13] That does not mean, of course, that the technology itself needs to be simple. Quite the contrary. It often takes highly complex (but robust and special-purpose) technology to create a device, which can simply be taken for granted by the user in pursuing her goals and projects. What matters is that as far as our conscious awareness is concerned, the tool itself fades into the background, becoming transparent in skilled use. In this respect the technology becomes, to coin a phrase "pseudo-neural." In childhood we learn how to use our various neural circuits to guide actions (learning to read, to walk, to talk, to write) and, later on, we simply take those capacities for granted as we confront the problems of adult life (preparing the business presentation, going out for groceries, etc.).

Personal information appliances, functioning robustly, transparently, and constantly, will slowly usher in new social, cultural, educational, and institutional structures. Perhaps we will one day live in a world in which, thanks to some easy-to-access implant or wearable device, your answer to the clepsydra question—like the one about the time—is simply, "Yes, (tiny delay) it means 'water clock,'" rather than "No, hold on while I go and look it up." For our sense of self, of what we know and of who and what we are, is surprisingly plastic and reflects not some rigid preset biological boundary so much as our ongoing experience of thinking, reasoning, and acting within whatever potent web of technology and cognitive scaffolding we happen currently to inhabit.

That web is already beginning to include a varied and mutually empowering matrix of human-centered technologies. Other elements in the near-future matrix include the development of lightweight, constantly running, personal computing appliances and of new techniques for rendering the informational substantial, thus blurring the boundaries between the virtual and the physical ("tangible computing," more on which below).

To see why we need this even-richer web of support, reflect that the original vision of Ubiquitous Computing, with its image of a smart world populated by semi-intelligent desks, doors, freezers, and coffeemakers, aims to put all the computational work out of sight. It seems unlikely, however, that we will do away with all need for personal data storage and knowledge access. As a result, we cannot really off-load **all** the computational work onto some fixed environment. Some of it should be, as Norman realized, linked more directly to a specific user. Wearable Computing, by attaching (quite literally) certain resources directly to the biological agent, offers a nonpenetrative means of catering to just this need. Instead of seeing Wearable and Ubiquitous Computing as competing approaches, then, it is much more fruitful to consider their large potential for harmonious interaction.[14]

A Wearable Computer is an information-processing tool that is, in a deep but noninvasive sense, integral to the user. It is portable, constantly running, and may be used while the agent is in motion or otherwise engaged. As such, it should support hands-free use and be capable of presenting data unobtrusively to the user whenever it sees fit. Such devices are "designed to be useable at any time with the minimum amount of cost or distraction from the wearer's primary task [which is] not using the computer [but] dealing with the environment."[15] Wearable Computing is thus, in a very broad sense, another instance of what Norman called a human-centered technology; it belongs, or aims to belong, to that species of technology that fades into the background in use, providing support while allowing us to focus not on the technology but on the task at hand. But it achieves this not by embedding the computing into the world but by affixing it to the agent.

An early example is Bradley Rhodes's "wearable remembrance agent."[16] This is described as "a continuously running proactive memory aid." The device comprises a commercially manufactured heads-up display, which presents an 80 × 25-character screen in the upper visual field via a slightly clumsy "hat-top" mounting.[17] More discreet technologies, including EyeGlass displays and laser-based retinal displays, could also be used.[18] The Microvision firm has piloted a device that uses safe laser technologies to scan images directly onto the user's retina.[19] Ultimately, one can imagine direct electronic input into V1, the main visual processing gateway to the brain, somewhat along the lines of the cochlear implants described in chapter 1. In Rhodes's device the heads-up display is combined with a special one-handed keyboard for input, known as a Twiddler keyboard (made by Handykey in New York) and a special software package, the Remembrance Agent (RA) itself. The software is designed to run constantly, and to respond to inputs by intelligently searching through the agent's local or distal file spaces for items whose contents match the current probe. Think Google, but imagine the resource roving over your own personal file spaces, looking for the notes you yourself entered last time you encountered such-and-such a person or situation. In principle, search-initiating probes could also originate from eyeglass-mounted cameras linked to face-recognition software and/or from signals continuously broadcast by local devices (here is one potential source of synergy with Ubiquitous Computing approaches). A typical pattern of use, as imagined by the designer, might go like this:

> Say the wearer of the RA system is a student headed to a history class. When she enters the classroom, note files that had previously been entered in that same classroom at the same time of day will start to appear . . . when she starts to take notes on Egyptian hieroglyphics, the text of her notes will trigger suggestions pointing to other readings and note files . . . when she later gets out of class and runs into a fellow student, the identity of the student is either entered explicitly or conveyed through an active badge system or automatic face recognition. The RA starts to bring up suggestions pointing to notes entered while around this person, including an idea for a project proposal that both students were working on. Finally, the internal clock of the wearable gets close to the time of a calendar entry reminding the wearer of a meeting . . . [20]

The idea is thus to combine the advantages of personal, agent-specific information, storage, and retrieval with input from a variety of fixed, environmentally distributed resources providing the wearable device with a stream of useful context-fixing information, helping it to guess where the agent is and what she is probably doing. The user is at once a mobile locus of highly personalized resources and a useful interface for local, embedded computational devices. She is also a kind of automatic electronic trail-leaver, whose movements and choices can be tracked—for good or ill—by the devices she passes near (for much more on this, see chapters 6 and 7). Wearable Computing and Ubiquitous Computing are natural allies whose full synergistic potential has yet to be explored.

 There is, however, another problem lurking in the general move toward ever-more-integrated, invisible, automatic, pseudo-neural technologies. The danger is one of loss of control.[21] Opaque technologies were, of course, hard to use and control; that's what made them opaque. But truly invisible, seamless, constantly running technologies resist control in a subtler, perhaps even more dangerous, manner. How then can we alter and control that of which we are barely aware? Suppose, for example, I am unhappy with the performance of my biological memory regarding names and dates. There is nothing very direct that I can do about this. I might engage in memory-training exercises; I might augment my biological memory with new resources (Palm Pilot, remembrance agent); I might try some neurotropic substances like *Gingko biloba*. But—considered as a piece of "cognitive kit"—my biological memory is pretty hard to get at and reconfigure. It is too far along the spectrum that leads to fully invisible computing.

Proponents of what has become known as tangible computing take this kind of worry very seriously indeed. In making our technologies truly, permanently invisible to the user, we may similarly limit our own capacities for creative intervention. The philosopher Heidegger, writing in 1927, distinguished between a tool's being "ready-to-hand" and its being "present-at-hand." The hammer, while in use, is ready-to-hand. It is not an object of conscious reflection. We can, in effect, "see right through it," concentrating only on the task (nailing the picture to the wall). But, if things start to go wrong, we are still able to focus on the hammer, encountering it now as present-at-hand, that is, as an object in its own right. We may inspect it, try using it in a new way, swap it for one with a smaller head, and so on. The effective use of tools thus often involves a kind of flipping between invisibility-in-use and availability for thought and inspection. Paul Dourish, a leading proponent of tangible computing, thus reminds us that "the effective use of tools inherently involves a continual process of engagement, separation and re-engagement."[22] Dourish, a one-time colleague of Mark Weiser, invented the term "tangible computing." Tangible computing maintains key elements of the invisible computation model but seeks to do so without allowing the tools and technologies to become permanently invisible, available solely as ready-at-hand. In common with the work discussed earlier, however, the interactions between the user and the tool are meant to be as natural and easy as possible, and to make the most of our basic skills and knowledge.

An appealing example of tangible computing is the Marble Answering Machine designed by Durrell Bishop at London's Royal College of Art.[23]

Standard digital answering machines can be hard to use and control; they may have multiple functions hidden in menus and useable only via complex sequences of button-pushings and key-holdings. By way of contrast, Bishop's design, Dourish tells us, works like this:

> [The] answering machine has a stock of marbles. Whenever a caller leaves a message . . . it associates that message with a marble from the stock, and the marble rolls down a track to the bottom, where it sits along with the marbles representing previous messages. When the owner of the machine comes home, a glance at the track shows . . . how many messages are waiting: the number of marbles arrayed at the bottom of the track. To play a message, the owner picks up one of the marbles and drops it in a depression at the top of the answering machine; because each marble is associated with a particular message, it knows which message to play. Once the message has been played, the owner can decide what to do: either return the marble to the common stock for reuse (so deleting the message) or returning it to the track (saving it to play again later).[24]

Now imagine a good spy-movie scenario. You receive a long and vital message on the home machine, but you have just one minute until your nosy roommate gets home. You don't want her to see or hear the message, but you cannot delete it yet. What do you do? I am willing to bet that every single reader of this text immediately thought "Take the marble and put it in your pocket, then later, in private, drop it into the machine." The situation was unusual, yet you at once knew how to proceed—but what would you have done on your usual digital device?

The point is that the Marble Answering Machine, by giving a familiar kind of physical presence to what is really a digital abstraction (the message), allows us to use our well-developed intuitions about physical objects to interact with the virtual/informational realm. As Dourish explains, the problem of interacting with the virtual is thus transformed into the more familiar one of interacting with concrete, movable objects. Instead of pushing technology to become totally invisible, the idea is to make it extravisible: to take digital abstractions and data-flows and make them as solid and manipulable as rocks and stones. In so doing, it is hoped, we provide for the kind of easy flippability (between ready-to-hand and present-at-hand) characteristic of many of our favorite tools.

Our biologically given neural "tools" typically lack this characteristic flippability, and it may be thought that this is sufficient reason to reject the idea of a potential symmetry between neural structures and any non-biological equipment that is *ever* encountered as an object in its own right. This cannot be quite right, however, since we can easily imagine encountering some of our own neural mechanisms as objects too. Biofeedback techniques already allow an indirect form of this, as when we learn (by way of audible signals) to induce neural alpha rhythms at will, to lower our blood pressure, and so on. More direct forms of encounter may become commonplace as neuro-imaging techniques allow us to watch our own brains as they process information. In so doing, we surely do not render these aspects of our neural functioning less part of ourselves. The increasing visibility of our biological information-processing routines is, however, an interesting counterpoint to the increasing transparency of our best nonbiological props and aids.[25]

The Tangible Media Group at MIT Media Lab is also in pursuit of this vision of embodied digitality. Their goal is to create a new generation of interfaces that increasingly blur the distinction between the virtual/informational and the tangible/physical. A typical project is the aptly named Sensetable, a tabletop display that uses electromagnetic sensing to determine the position of a variety of physical objects (placed on the tabletop), which the user can then move around so as to amend and alter the information displayed.[26] One example is a chemistry teaching package in which a variety of atoms and molecules are displayed on the desktop screen. On top of the screen (i.e., on the desktop) there is also an array of small, puck-like objects. Each puck, if moved on top of a specific atom or molecule, becomes "bound" to that item (much as a marble became bound to a specific message). The puck is now a physical embodiment of that information. Moving the puck around on the display, or bringing it into contact with another puck, now causes the system to simulate the effects of bringing that atom or molecules into contact with others. Chemical reactions can be investigated, and new molecules built and examined. In addition, modifiers can be attached to the puck via a surface slot, in order to change the charge on that atom or molecule, and so on.

Sensetable is a descendent of a system called metaDesk, which used cameras and computer vision techniques (instead of electromagnetic sensing) to allow a variety of physical icons ("phicons") to interact with a table-top display.[27] Common to all these projects, then, is the use of what the group calls Tangible User Interfaces (TUIs) in which familiar physical objects, instruments, surfaces, and spaces are used to mediate our exchanges with digital information systems. Such interfaces aim, also, to erode the gap between input systems and output systems.[28] When I write on a piece of paper, the input space and the output space are one and the same; the stored item appears exactly where it was *input*. Standard Graphical User Interfaces (GUIs) pull the spaces apart: you type on a keyboard and the information is stored somewhere else, and displayed on a screen. TUIs use displays, which are themselves "aware" of the user's activities, and which act as input to the system (the puck-sensitive tabletop screen, for example).

One promising idea is to exploit the kinds of interface we find familiar in the noncomputational world to better mediate our contact with digital and informational resources. In this vein, Neil Gershenfeld and his colleagues produced a bow-using interface to mediate the contact between a world-class cello player (Yo-Yo Ma) and an electronic cello. The bow provides a superbly sensitive, delicately nuanced, feedback-friendly means of continuously controlling the musical ebb and flow. It is an interface that has been tuned and adapted over centuries of use, and to which the human cellist has devoted a lifetime of study. Why throw all that away in favor of a few buttons and a mouse? If synthesized music can sometimes sound cold and lifeless, might that have more to do with the use of such stale interfaces, rather than the potential of the digital medium itself? Recently, the bow-based interface was used to great effect by Yo-Yo Ma in a Tokyo performance. The digital media allowed the artist to create new sound combinations beyond the reach of any normal cello, while the familiar interface allowed him to explore these new possibilities with all his characteristic flair and insight. Ma himself, according to Gershenfeld, is engagingly enthusiastic and unsentimental, treating both his original cello (a Stradivarius) and the new array as just two technologies: simply the means to his musical ends.

Another area in which the notion of the interface is being reinvented is in work on Augmented Reality. In this work, the interface is nothing more than your own view of the world as you look around, but the view is augmented using some kind of heads-up or eyeglass style display system. The display might use video systems to mix computer graphics and input from cameras aimed at the scene before you, or take direct optical input and overlay it with computer graphics. The idea, in each case, is to overlay our experience of the physical world with layers of personalized digital information. This kind of work uses many of the same techniques and technologies as work on Virtual Reality and Wearable Computing. But instead of trying, as with standard Virtual Reality approaches, to re-create a simulacrum of the real physical world entirely *inside* some computer-generated realm, the goal of Augmented Reality is to *add* digital information to the everyday scene. Think of it as a kind of digital annotation and enhancement regime, with the specific annotations and enhancements being tailored to the needs and desires of different users passing through the (real-world) scene.

For example, combining Global Positioning information with locally poised digital resources makes it possible to associate specific items of information with geographical locations. Such information could be picked up using special eyeglasses, or via handheld or other wearable devices. Thus imagine you are lost on a university campus.[29] To find the library, you simply enter the name "library" in a handheld local guide-box and don a pair of special eyeglasses. As you look around, you see a giant green arrow take shape in the sky, pointing at the roof of the library! Looking down at the path, you see smaller arrows indicating the best route. Hanging in the air around your body you notice a variety of small icons offering you other local services. To use them, you just reach out and "touch" them, sending position and motion information through sensors in your clothing.

The term "Augmented Reality" was first used by a group of Boeing engineers and scientists in the early 1990s.[30] Their idea was to use such systems to help workers install complex wiring harnesses in aircraft. The workers would see the desired positioning superimposed upon the actual physical structure of the plane. In a similar vein, engineers seeking to repair broken equipment might soon see the innards of the machine alongside specific repair instructions highlighting the elements to be removed and replaced. Surgeons seeking to repair human brains or bodies could benefit in the same way, seeing ultrasound scans or brain imaging information projected onto the appropriate areas. Researchers at the University of Central Florida have overlaid a model of a knee-joint on a woman's leg.

Using infrared LEDs (Liquid Electronic Displays) to inform the system about current leg position, the Augmented Reality interface allows onlookers to see just how the bones move while the woman walks and bends. The use of overlaid digital resources to enhance our ordinary daily experience of the world and to provide new means of physical-virtual interaction is likely to play a major role in the next decade. Very soon we may expect to see various kinds of personalized electronically overlaid information, from advertising to information about incoming cell phone calls or even about ourselves apparently suspended in the air as we roam about. Such information might appear attached to the space around an individual, or a shop, or a designated electronic advertising area. Once again, the key innovation is to allow the physical and the informational realms to seamlessly merge and mingle, in ways that unobtrusively support daily activity and that make maximum use of our normal means of embodied, socially embedded activity. It is worth repeating that such work in Augmented Reality, though it uses some of the same technologies as Virtual Reality, is really quite different at root. The aim here is not to create a richly detailed version of the daily world inside the machine, but to use the machine to add new layers of meaning and functionality to the daily world itself. Some theorists thus speak not of Virtual Reality but of Real Virtuality—a kind of deliberate blurring of the boundaries between physical and informational space.

This kind of blurring has educational importance too. If we are indeed becoming complex biotechnological hybrids, a major challenge for the future will be to train young minds to think well about a world in which the physical and the informational/digital are densely and continuously interwoven. To that end, researchers are developing forms of so-called mixed reality play.[31] In mixed reality play, the virtual/informational is made tangible, the physical made virtual, and the two realms interwoven in single play-based experiences. One of the keys to such experience is the development of what the team calls "traversable interfaces between real and virtual worlds." A traversable interface creates the impression of a seamless join between the real and the virtual, and encourages users to frequently and naturally cross over between the two realms.

Versions of such techniques can allow children to engage in "mixed reality play," in which a coherent play space is created with characters who are able to cross over between the physical and virtual realms. In one of the last papers written before his untimely death, my colleague and friend Mike Scaife, working as part of a multi-university interdisciplinary research collaboration, helped design a mixed reality adventure game called "Hunting the Snark."[32] The game targets children in the 6 to 10 age group, and the idea is to locate an elusive entity (the Snark) that can live in both the physical and the digital world, and whose activities seamlessly crisscross the two realms. This game also relies on traversable interfaces that give the illusion of joining the two worlds. For example, physical items bearing electronic tags can be used to trigger events in the digital world, when placed into some key location (e.g., a kind of magic well) in the physical world. Such key locations included the aforementioned well, a wardrobe, a cave, and a mouth. When placing a real object in the well, the children could see the object in their hand disappear in the real world, emerging at once in the digital one, only to be eaten, or rejected, by the digital manifestation of the Snark (figs. 2.1 and 2.2). The use of simple forms of wearable computing also allowed some of the children's real movements to be coordinated with the movements of the digital Snark. This kind of playful technology provides novel and exciting experiences, which should help young brains learn better how to make the most of a world in which the physical and the digital are ever-more-closely intertwined—worlds in which everyday objects (medicine cabinets, coffeemakers, refrigerators) have informational state, and informational phenomena have much more tangible physical presence.

Nurtured by such experiences, and living and moving in a world populated with ubiquitous computing devices, augmented reality displays, and various kinds of tangible computing, next-generation human minds will not invest very heavily in the virtual/physical divide. Instead, these minds

---

(*Facing page, top*)   **Fig. 2.1**   The Snark, a being that straddles the digital and physical worlds, displays a happy expression when the children place suitable, electronically tagged objects in the well. Such "mixed reality play" may help young brains learn how to think better about a world in which the digital and the physical are ever-more-closely intertwined. Image courtesy of Eric Harris and Yvonne Rogers.

(*Facing page, bottom*)   **Fig. 2.2**   The Snark can look unhappy when hungry or presented with the wrong objects. Image courtesy of Eric Harris and Yvonne Rogers.

will focus on activity and engagement, seeing both the virtual and the physical as interpenetrating arenas for motion, perception and action.[33] Mixed reality play intends to block the stale opposition between the real and the virtual, or the bodily and the informational, revealing each for what it is: just one more aspect of a larger world in which hybrid selves live, move, work, and play.

Moving On

Time to take stock. We have now met several visions of the near future. The visions of Invisible Computing, of Tangible Computing, of Wearable Computing, and of Augmented Reality. Of these, Invisible Computing and Tangible Computing at first seem like diametrically opposed research programs, but this is not really the case. The differences are real, but easily overplayed. Is the wristwatch an example of invisible or tangible technology? Norman and Weiser's original vision does not require total invisibility so much as invisibility-in-use, and on this, both models converge. The Marble Answering Machine, to take another case, is every bit as good an exemplar of an Information Appliance as it is an instance of Tangible Computing. And it is the very same features that make it "tangible" (the way it exploits our ease and familiarity with everyday objects) that allow it to become invisible in daily use.

The differences between the two visions thus show up only, if at all, at the very extremes, where some Information Appliances will indeed be designed to remain firmly out of sight and out of mind. But sometimes, as Norman stresses and Dourish would surely admit, this is the best way. The engine management system of a modern automobile is a case in point, and the intelligent coffeepots and carports imagined by Norman and Weiser may be others. The question we really should *not* ask may be, Which way is best? That is rather like asking whether our best tools should be more like hands, hammers, or the hippocampus. The question is misguided, because each of these tools is specialized for different purposes and (hence) needs to be accessed, used and/or reconfigured in very different ways. For certain purposes we want tools that we can step back from and think about. For other purposes, we want tools that function continuously and quasi-independently, requiring little or no conscious attention and that resist easy reprogramming (more like the homeostatic control systems that regulate heart rate, breathing, and the like discussed in chapter 1).

The various kinds of transparent, human-centered technologies that we have so far imagined are, however, typically restricted in one specific way. All the *fitting*, the adaptation of the technology to the needs and capacities of the biological user, is done by the slow cultural process of design and re-design; the final dovetailing of biology and technology is achieved courtesy of individual human learning. This neglects the important—perhaps ulti-mately transformative—potential of information appliances that, in use, actively work to learn about and better fit the user. I dub such appliances "dynamic appliances." Not all dynamic appliances are transparent and unobtrusive. Existing speech-to-text software is quite hard to set up and use, but it is dynamic in that it learns about specific users and adapts to *their* voices and vocabularies. The combination of dynamic appliances and transparent technologies is surely a match made in cyborg heaven. Imagine information appliances that actively learn about the user. First, you take the word-processing function out of the PC and lodge it in a special, dedi-cated writing and composing platform—an information appliance using a nice Direct Manipulation Interface. Next, you allow the machine to monitor its own use. After a while, unused functions can be temporarily disabled, and frequently used functions are given a more efficient implementation. Response times speed up. The appliance has become *skilled* at doing just what the user requires, no more and no less. This is what we already find in our own neural structures, and pseudo-neural technologies need to aim for the same kind of effect. There is a cost of course: the *immediate* functional-ity is reduced. But this is a trade-off the biological brain makes all the time. You become skilled at driving a car with a certain configuration of controls. After a while, the cognitive effort needed to drive that kind of car drops, and the capacity to rapidly and effectively avoid collisions increases. The cost of this is felt if driving demands change—say you rent a car with a very alien configuration. But the process then repeats. The biological brain is constantly striving to streamline, chunk, compile, and automate, and it does so by attending to repeated patterns of activity and use. Dynamic information appliances would, when appropriate, do just the same. The combination of brains that learn about technologies, with ubiquitous, in-creasingly transparent technologies that "learn" about individual brains, sets the scene for a cognitive symbiosis whose full potential and implications none of us can yet fully appreciate.

The technological present, then, is a shifting kaleidoscope of visions of the future. The smart world full of invisible technologies; the world of constantly running, easily deployed wearable computers; the world of neuro-electronic implants; the world of tangible computing and real virtuality; the world of dynamic, self-reconfiguring wearables and information appliances. Which is it to be? The visions jostle for space, but they are not truly competing. Quite the reverse. These are, as we will see, complementary threads in an emerging biotechnological fabric. Our cyborg future, like our cyborg present and our cyborg past, will depend on a variety of tools, techniques, practices, and innovations. What they will increasingly have in common is that deep human-centeredness that Norman so powerfully celebrates. These will be technologies to live with, to work with, and to *think through*. Such technologies are apt for the most profound and enduring kinds of interweaving into our lives, identities, and projects, and into our constantly constructed sense of place, presence, and self.

CHAPTER 3

Plastic Brains, Hybrid Minds

Your own body is a phantom, one that your brain has temporarily constructed purely for convenience.

—V. S. Ramachandran and S. Blakeslee

The Negotiable Body

Here are some playful—but important and illuminating—experiments you
can do at home. They were designed by V. S. Ramachandran, who is professor and director of the Center for Brain and Cognition at the University of
California, San Diego.[1] Follow the simple instructions and you will (with
about 50 percent probability) feel as if your nose is two feet long, feel as if
the desktop is part of you and capable of feeling pain, and feel sensation in
a dummy (rubber) hand.

The point of these experiments is to show that our sense of our own
bodily limits and bodily presence is not fixed and immovable. Instead, it is
an ongoing construct, open to rapid influence by tricks and (as we'll see in
chapters 4 and 5) by new technologies.

Experiment One: The Extended Nose

Arrange two chairs in a line, one behind the other. Seat yourself in the rearmost
chair and have a friend blindfold you. Get a volunteer to sit in the chair in
front of you. Now get your friend to stand beside the two occupied chairs and
issue the following instructions, taken from Ramachandran and Blakeslee:

Take my right hand and guide my index finger to [the seated volunteer's] nose. Move my hand in a rhythmic manner so that my index finger repeatedly strokes or taps [the volunteer's] nose in a random sequence like a Morse code. At the same time, use your left hand to stroke my nose with the same rhythm and timing. The stroking and tapping of my nose and [the volunteer's] nose should be in perfect synchrony.[2]

After less than a minute of this synchronized nose-tapping, about half the subjects report a powerful illusion. It is as if their own noses now extended about two feet in front of them. Here's why: your brain registers the rhythmic tapping of your finger and knows that your arm is extended out in front of you. It is also receiving signals, perfectly coordinated with this tapping routine, from the end of your own nose (the friend is tapping your nose in synchrony with the tapping of your finger on the volunteer's nose). To make sense of this close and ongoing match between arm's-length tapping and end-of-nose sensation, the brain infers that your nose must now extend far enough for the arm's-length tapping to be causing the feelings. So your nose *must* be about two feet long. So that's how it (suddenly) feels to you. But this, as Ramachandran and Blakeslee go on to comment, is really quite extraordinary: "Your certain knowledge . . . constructed over a lifetime [can be] negated by just a few seconds of the right kind of sensory stimulation."

Experiment Two: A Pain in the . . . Desktop?

Sitting at your desk, place your left hand underneath the desktop. Get a volunteer to tap the desktop with her right hand while using the left to (in synchrony) tap your hidden hand. Once again, many subjects will feel as if the "being tapped sensation" is located on the desk surface—as if the desktop were a real, sensitive part of their body. Now have the volunteer hit the desktop with a hammer. Your galvanic skin response jumps as if your own hand had been threatened![3]

Experiment Three: Sensation in a Dummy Hand

A variant of the last experiment uses a plastic dummy hand. A partition is created so that you see only the dummy hand, and a volunteer again taps both your real hand (hidden behind a screen) and the dummy hand (in your direct view) in perfect synchrony. Subjects experience sensations "in the dummy hand."

In all these cases (and you can probably now dream up many more), we discover that the body-image supported by a biological brain is quite plastic, and highly (and rapidly) responsive to coordinated signals from the environment. The image of the physical body with which we so readily align our pains and pleasures is highly negotiable. It is a mental construct, open to continual renewal and reconfiguration.

One reason this makes sense, of course, is that our bodies *do* change during our lifetimes. Limbs grow and develop and sometimes are tragically lost. Ramachandran himself has worked extensively with so-called "phantom limb" patients, and it was this work that, in part, led him to devise and carry out the experiments just rehearsed. Phantom limb patients (amputees who continue to feel either motion, movement, or pain in the missing limb), he found, could sometimes be helped by devising ways to fool their brains into thinking the missing limb was in various states. For example, a patient with sensations as of a clenched and painful phantom hand was helped with a box and mirror arrangement allowing an image of the real remaining hand to be cast into the space "occupied" by the phantom.[4] The patient could then relieve the pain in the phantom limb simply by clenching, then unclenching, the real hand. The visual feedback of unclenching occurring in a hand that seemed (courtesy of the mirror box) to be located where the phantom should be, allowed the brain to re-organize its body-image so as to eliminate the sensation of clenching. The trick was thus to create a good enough "virtual reality," in which the phantom limb was visually available and responsive to the patient's intentional control and manipulation. This allowed the re-negotiation of a (less painful) orientation for the phantom limb. The key idea, common to both this "mirror box" work and to the experiments rehearsed earlier, is that *despite* the probable presence of some preset genetic components in our body-images, there is also—and simultaneously—large scope for continual revision.[5] The deeper principle underlying all such revisions and re-negotiations now looks reasonably clear-cut. It is that our brains depend on *perceived correlations* (for instance, the correlation between observed desk-tappings and felt sensation) to continuously construct a model of—and hence a sense of—our bodily bounds and locations. We can call this Ramachandran's principle. In Ramachandran and Blakeslee's own words,

For your entire life, you've been walking around assuming that your "self" is anchored to a single body that remains stable and permanent at least until death. . . .Yet these [results] suggest the exact opposite—that your body image, despite all its appearance of durability, is an entirely transitory internal construct that can be profoundly altered with just a few simple tricks.[6]

The implications of this discovery for technology-based manipulations of our sense of presence, body, and location are enormous, as we shall see in chapter 4.

To recap, human brains (and indeed those of many other animals) seem to support highly negotiable body-images. As a result, our brains can quite readily project feeling and sensation beyond the biological shell. In much the same way, the blind person's cane or the sports star's racket soon come to feel like genuine extensions of the user's body and senses. Once again, this is because our continual experience of closely correlated action and feedback routines running via these nonbiological peripheries allows the brain to temporarily generate what is really a new kind of "body-image," one that includes the nonbiological components. The transformative effects of this run pretty deep. In a recent neuroscientific experiment in which a monkey repeated food-retrieving actions using a rake, experimenters reported that

the visual RF [receptive fields] of cells in the anterior bank of the intraparietal sulcus became elongated along an axis of the tool, as if the image of the tool was incorporated into that of the hand.[7]

Otherwise put, the monkey's brain rapidly learned to quite literally treat the rake as an extension of its fingers. It is reasonable to suspect that it is at precisely this point that certain kinds of tools (manually deployed ones) become transparent in use. Here, as elsewhere, the seeds of the most intimate organism-artifact unions are sown by the biological brain itself.

Neural Opportunism

Several other features of our brains combine to make us humans especially open to processes of deep biotechnological symbiosis. One such feature is what I'll call "neural opportunism." Sit back in your chair and take a look around the room. What did you see? In all likelihood, you had the experience of a succession of rich visual images: images of chairs, books, tables, CDs, audio equipment, whatever. In my own case, I saw a bookshelf stacked rather untidily with too many things I ought to have read, their multicolored spines accusingly flaunting clear, crisp, inviting titles. Looking around, I glimpsed an open closet liberally sprinkled with gaudy Hawaiian shirts, stark against the mundane backdrop of darker, workaday clothing. But now let's ask what turns out to be an especially tricky question. In generating that sequence of visual experiences, what information did my biological brain actually bother to extract and process? The answer is—significantly less than we might have guessed.

To understand this, first reflect that the human visual system supports only a small area of high-resolution processing, corresponding to the fraction of the visual field that falls into central focus. When we inspect a visual scene, our brains actively move this small high-resolution window (the fovea) around the scene, alighting first on one location, then another. The whole of my bookcase, for example, cannot possibly fit into this small foveal area while I remain seated at my desk. My overall visual field (that area *plus* the low-resolution peripheries) is, of course, much larger, and a sizable chunk of my bookshelf falls within this coarse-grained view. It has been known since 1967 that the brain makes very intelligent use of its small high-resolution fovea, moving it around the scene (in a sequence of rapid motions known as visual saccades) in ways delicately suited to the specific problem at hand.[8] This can be seen from the fact that human subjects presented with *identical* pictures, but told to prepare to solve *different* kinds of problems (some might be told to "give the sex and ages of the people in the picture," while others are asked to simply "describe what is going on" and still others to prepare to "recall the objects in the room"), show very different patterns of visual saccade. These saccades, it is also worth commenting, are fast—perhaps three per second—and often repetitive, in that they may visit and revisit the very same part of the scene. What are they for?

One possibility, at this point, is that each saccade is being used to slowly build up a detailed internal representation of the *salient* aspects of the scene. The visual system would thus be selective, but would still be doing what we intuitively expect. It would be using visual input to slowly build up a detailed neural image of the scene. Subsequent research, however, suggests that the real story is even stranger than that. We can get a sense of this even before looking at the scientific experiments, by thinking about some magic tricks.

There is an entertaining web site where you can try out the following trick.[9] You are shown, on screen, a display of six playing cards (new ones are generated each time the trick is run). In the time-honored tradition, you are then asked to mentally select and remember one of those cards. You click on an icon and the cards disappear, to be replaced by a brief "distracter" display. Click again and a five-card (one less) array appears. As if by magic, the very card that you picked is the one that has been removed. How can it be? Could the computer have somehow monitored your eye movements? A version of this trick is displayed on pages 65 and 66 of this book. Go to page 65 and immediately pick a card from the display shown in Fig. 3.1. Concentrate on that card. Remember it. Now go to page 66. Did we remove the very card you chose? Amazing isn't it! I must confess that on first showing (and second, and third) I was quite unable to see how the trick was turned.

Here's the secret. The original array will always show six cards of a similar broad type: six face cards, or six assorted low-ranking cards (between about two and six, for example). When the new, five-card array appears, NONE of these cards will be in the set. But the new five-card array will be of the same type: all face cards, low cards, whatever. In this way, the trick capitalizes on the visual brain's laziness (or efficiency, if you prefer). It seems to the subject exactly as if all that has happened is that one card (the very one he mentally selected!) has disappeared from an otherwise unchanged array. But the impression that the original array is still present is a mistake, rooted in the fact that all the brain had actually encoded was something like "lots of royal cards including my mentally selected king of hearts." Magic tricks such as these rely on our tendency to overestimate what we actually see in a single glance, and on the manipulation of our attention so as to actively inhibit the extraction of crucial information at certain critical moments. The philosopher Daniel Dennett makes a similar point using a different card trick.[10] He invites someone to stand in front of him and fixate on his (Dennett's) nose. In each outstretched arm he holds

Fig. 3.1 Pick a card and concentrate on it very hard. We will make your card, and only your card, disappear. Turn to page 66 and see if your card is now missing from the array! Thanks to Andy Bauch for permission to show the trick here.

a playing card. He brings his arms in steadily. The question is, at what point will the subject be able to identify the color of the card? Here too, we may be surprised. Color sensitivity, it turns out, is available only in a small and quite central part of the visual field. Yet my conscious experience, clearly, is not of a small central pool of color surrounded by a vague and out-of-focus expanse of halftones. Things look colored all the way out. Once again, it begins to look as if my conscious visual experience is overestimating the amount and quality of information it makes available.

Now imagine that you are the subject of another famous experiment.[11] You are seated in front of a computer screen on which is displayed a page of text. Your eye movements are being automatically tracked and monitored. Your experience, as you report it, is of a solid, stable page of readable text. The experimenter then reveals the trick. In fact, the text to the left and right of a moving "window" has been constantly filled with junk characters, not recognizable English text at all. But because that small window of normal, readable text has been marching in step with your central perceptual span, you never noticed anything odd or unusual. For comparison, this is as if my bookshelf only ever once contained (at the same moment) four or five clearly titled books, and the rest of the titles were all senseless junk. Nonetheless, it would have looked to me as if I were seeing a wide array of clear English titles at all times. In the case of the screen of text, the window of "good stuff" needed to support the illusion is about eighteen characters wide, with the bulk of the characters falling to the right of the point of fixation (probably because English is read left to right).

Similar experiments have been performed using pictures of a visual scene, such as a house, with a parked car and a garden.[12] As before, the victim sits

Fig. 3.2 Did we get your card? Puzzled? Go back to page 65 and try again.

in front of a computer-generated display. Her eye movements are monitored and, while they saccade around the display, changes are clandestinely made: the colors of flowers and cars are altered, the structure of the house may be changed; yet these changes, likewise, go undetected. We now begin to understand why the patterns of saccade are not cumulative—why we visit and repeatedly *revisit* the same locations. It is because our brains just don't bother to create rich inner models. Why should they? The world itself is still there, a complex and perfect store of all that data, nicely poised for swift retrieval as and when needed by the simple expedient of visual saccade to a selected location. The kind of knowledge that counts, it begins to seem, is not detailed *knowledge* of what's out there, so much as a broad *idea* of what's out there: one capable of then informing on-the-spot processes of information retrieval and use.

Finally, lest you suspect that these effects (known as "change blindness") are somehow caused by the unnaturalness of the experimental situations, consider some recent work by Dan Simons and Dan Levin.[13] Simons and Levin took this research into the real world. They set up a kind of slapstick scenario in which an experimenter would pretend to be lost on the Cornell campus, and would approach an unsuspecting passerby to ask for directions. Once the passerby started to reply, two people carrying a large door would (rudely!) walk right between the inquirer and the passerby. During the walk through, however, the original inquirer is deftly replaced (under cover of the door) by a different person. Only 50 percent of the subjects (the direction-givers) noticed the change. Yet the two experimenters were of different heights, wore different clothes, had very different voices, and so on. The conclusion that Simons and Levin draw is that our failures to detect change are not due to the artificialness of the computer-screen experiments. Instead, they arise because "we lack a precise representation of our visual world from one view to the next" and encode only a kind of rough gist of the current scene—enough to support a broad underlying sense of what's going on *insofar as it matters to us*, and enough to guide further intelligent information-retrieval, via directed saccades, as and when needed.[14]

A final demonstration of these startling effects can be obtained using the so-called flicker paradigm. Here, you look at a computer-generated image, which flashes on and off, with a masking screen intervening. Between each showing of the image, something changes. Even when these changes are large and significant (for example, one jet engine of an airplane, shown at center screen, repeatedly appears and then disappears), we do not easily spot them. For many of these changes, subjects need to view the rapidly alternating images for nearly a minute before they see the change. Once they have spotted the change they find it hard to believe that they did not see it at once. Normally, motion cues would alert us to the area of the visual scene where a change was occurring. But in these experiments the motion cue is being screened off by the intervening blank screen (the mask). Without that cue, the changes prove very hard to detect. You can try these experiments out at various sites on the web listed in the note.[15]

What all this suggests is that the visual brain may have hit upon a very potent problem-solving strategy, one that we have already encountered in other areas of human thoughts and reason. It is the strategy of preferring *meta-knowledge* over *baseline knowledge*. Meta-knowledge is knowledge about how to acquire and exploit information, rather than basic knowledge about the world. It is not knowing so much as knowing how to find out. The distinction is real, but the effect is often identical. Having a super-rich, stable inner model of the scene could enable you to answer certain questions rapidly and fluently, but so could knowing how to rapidly retrieve the very same information as soon as the question is posed. The latter route may at times be preferable since it reduces the load on biological memory itself. Moreover, our daily talk and practice often blurs the line between the two, as when we (quite properly) expect others to know what is right in front of their eyes. Or when—to recall an example from the previous chapter—we say that we know the time, before looking, simply because we are wearing a watch!

The visual brain is thus *opportunistic*, always ready to make do and mend, to get the most from what the world already presents rather than building whole inner cognitive routines from neural cloth. Instead of attempting to create, maintain, and update a rich inner representation (inner image or model) of the scene, it deploys a strategy that roboticist Rodney Brooks describes as "letting the world serve as its own best model."[16] Brooks's idea is that instead of tackling the alarmingly difficult problem of using input from a robot's sensors to build up a highly detailed, complex inner model of its local surroundings, a good robot should use sensing frugally in order to select and monitor just a few critical aspects of a situation, relying largely upon the persistent physical surroundings themselves to act as a kind of enduring, external data-store: an external "memory" available for sampling as needs dictate.

Our brains, like those of the mobile robots, try whenever possible to let the world serve as its own best model. In the light of this, some writers have suggested that our daily experience of a rich, highly detailed visual scene unfolding before the mind's eye must be something of an illusion.[17] On this view, it only *seems* to us as if we enjoy rich visual experience, thanks to that rapid capacity to retrieve more detailed information from the world as and when required. I now suspect, however, that this is a rather more delicate call than it at first appears, and the reason is one that bears quite directly on the larger themes of the present treatment.[18]

To see what I mean, let's leave the visual case (temporarily) and consider a very different example. Imagine you are a devout sports fan, and that you know thousands upon thousands of somewhat arcane facts about the performance statistics of players in U.S. women's basketball over the last twenty years. One day, as you are seated on your favorite barstool awaiting the start of a game, conversation turns to the Sacramento Monarchs' Kedra Holland-Corn. You immediately recall a few useful facts: that in 2000, her three-point field goal percentage was .361, ranking her seventeenth in the WNBA; that she scored a staggering twenty-three points in 8-of-12 shooting in a recent win over Los Angeles, and so on. While you reel off these facts and figures, you are implicitly aware that you could have done the same for any number of other players in the WNBA. You are not currently thinking about, for example, Jennifer Azzi of the Utah Starzz. But had the need arisen, her field throw percentage of .930 in the 2000 season would have been as readily available as the data on Holland-Corn. Hence, we have no hesitation in ascribing to you a rich underlying body of basketball knowledge. It is not that all that knowledge is currently *conscious*. You are not, let us imagine, right now experiencing any thoughts about Jennifer Azzi, only about Kedra Holland-Corn, but you do experience yourself as *in command* of a rich and detailed database in which all that information is stored, organized, and poised for easy recovery and use. Returning to the case of vision, notice that there, too, we find ourselves in command of a rich and detailed visual database in which information about the current scene is stored, organized, and poised for use. It is just that much of the database, in the case of vision, is located *outside* the head and is accessed by outward-looking sensory apparatus, principally the eyes. In each case, however, it is the fact that you can indeed access all this data swiftly and easily as and when required that bears out our judgments about the richness of our own knowledge and understanding.

Word Brains

You can probably see where this is heading and how it fits in with our emerging cyborg theme tune. It just *doesn't matter* whether the data are stored somewhere inside the biological organism or stored in the external world. What matters is how information is poised for retrieval and for immediate use as and when required. Often, of course, information stored outside the skull is not so efficiently poised for access and use as information stored in the head. And often, the biological brain is insufficiently aware of exactly *what* information is stored outside to make maximum use of it; old fashioned encyclopedias suffer from all these defects and several more besides. But the more these drawbacks are overcome, the less it seems to matter (scientifically or philosophically) exactly *where* various processes and data stores are physically located, and whether they are neurally or technologically realized. The opportunistic biological brain doesn't care. Nor—for many purposes—should we.

Consider next the opportunistic infant brain in the ecologically unique environment of spoken and written words. What might the reliable presence of linguistic surroundings do for brains like ours? This is a complex and much-debated issue.[19] But the small thread that I want to pull on here concerns the role of spoken language itself as a kind of triggering cognitive technology. Words, on this account, can be seen as problem-solving artifacts developed early in human history, and as the kind of seed-technology that helped the whole process of designer-environment creation get off the ground.

Let us bracket the difficult question of what, perhaps relatively small, biological changes and adaptations allowed the process of language-generation and understanding to get going in the first place. To do this is, of course, to bracket a lot.[20] To the contemporary infant brain, public language is simply encountered during early experience. The words for "car" and "drugstore," and indeed the practice of labeling cars and drugstores at all, are not things the normal child has to invent, any more than she has to invent parks, playgrounds, or playgroups. They are all simply aspects of the strange and highly structured world into which she is born.

Our question, then, is what occurs when opportunistic infant brains encounter the world of language? One thing that happens is that a variety of *cognitive shortcuts* become available, allowing brains like ours to explore and understand realms that would otherwise prove intractable or simply invisible. My favorite example of this comes from work not on humans but on a type of chimpanzee, *Pan troglodytes*. U.S.-based researchers Thompson, Oden, and Boysen trained chimps to associate a simple plastic token (such as a red triangle) with any pair of identical objects (two shoes, say) and a differently shaped plastic token with any pair of different objects (a cup and a shoe, or a banana and a rattle).[21] The token-trained chimps were subsequently able, without the continued use of the plastic tokens, to solve a more complex, abstract problem that baffled nontoken-trained chimps. The more abstract problem (which even we sometimes find initially difficult!) is to categorize *pairs-of-pairs* of objects in terms of *higher-order* sameness or difference. Thus the appropriate judgment for the pair-of-pairs "shoe/shoe and banana/shoe" is "different" because the *relations* exhibited within each pair are different. In shoe/shoe the (lower order) relation is "sameness"; in banana/shoe it is "difference." Hence the higher-order relation—the relation *between* the relations—is difference. By contrast, the two pairs "banana/banana and cup/cup" exhibit the higher-order relation "sameness," since the lower-level relation (sameness) is the same in each case. (See, I told you this wasn't easy!)

To recap, the chimps whose learning environments included plastic tokens for sameness and difference were able to solve a version of this rather slippery problem. Of the chimps not so trained, not a single one ever learned to solve the problem. The high-level, intuitively more *abstract*, domain of relations-between-relations is effectively invisible to their minds. How, then, does the token-training help the lucky (?) chimps whose early designer environments included plastic tokens and token-use training?

Here's what the experimenters suggest, and I find compelling. Imagine that the chimps' brains come to associate the sameness judgments with an inner image or trace of the external token itself. To be concrete, imagine the token was a red plastic triangle and that when they see two items that are the same they now activate an inner image of the red plastic triangle. Then imagine that they associate judgments of difference with another image or trace (an image of a yellow plastic square, say). Such associations reduce the tricky higher-level problems to lower-order ones defined not over the world but over the inner images of the plastic tokens. To see that "banana/ shoe" and "cup/apple" is an instance of higher-order sameness, all the brain now needs to do is recognize that two green triangles exhibit the lower-order relation sameness. The learning made possible through the initial loop into the world of stable, perceptible plastic tokens has allowed the brain to build circuits that reduce the higher-order problem to a lower-order one of a kind their brains are already capable of solving.

Notice, finally, that all that really matters to generate this effect is the association of the lower-order concepts (sameness and difference) with stable, perceptible items. Instead of plastic tokens, repeatable and distinctive *sounds* would have done just as well: a whistle for sameness, and a hum for difference. What, then, is the spoken language we all encounter as infants if not a rich and varied repository of such stable, repeatable auditory items? The human capacity for advanced, abstract reason owes an enormous amount to the way these words and labels act as a new domain of simple objects on which to target our more basic cognitive abilities. And the process is, of course, repeated. Given a simple label for second-order sameness, you and I could go on to make judgments about third-order sameness: sameness of higher-order sameness among pairs-of-pairs-of-pairs, and so on. After a while we might be forced to use pen and paper to keep track, thus falling back on yet another technological resource, but the trick remains the same. The whole imposing edifice of human science itself is testimony, I believe, to the power and scope of this species of cognitive shortcut. The simple act of labeling allows the biological brain to tiptoe into cognitive waters invisible, and hence impassable, to the languageless mind.[22]

Labels, of course, are not the whole story. The cultural tool of public language gives us not just labels but whole, structured, recursive *systems* for the encoding, objectification, and communication of thoughts and ideas. In learning such systems, the human brain is subjected to a potent and empowering dose of self-administered transformational medicine. It is not yet clear just how this all works, but the power is evident. Thus consider Joseph, a deaf eleven-year-old who was never taught sign language and had a childhood deprived of all structured linguistic experience. Here is a description from psychologist Oliver Sacks's 1989 book, *Seeing Voices*.

> Joseph saw, distinguished, categorized, used; he had no problems with perceptual categorization or generalization, but he could not, it seemed, go much beyond this, hold abstract ideas in mind, reflect, play, plan . . . he seemed, like an animal or an infant, to be stuck in the present, to be confined to literal and immediate perception.[23]

Returning to the impact of simple labeling, there is strong evidence that human mathematical abilities likewise seem to depend, in at least one crucial aspect, upon our experiences with the stable sound bites corresponding to individual number words. In an elegant series of investigations Stanislas Dehaene and colleagues have provided compelling evidence that precise numerical reasoning, involving numbers greater than three depends upon language-specific *representations of numbers*.[24] There is, to be sure, a kind of low grade, approximate numerical sensibility that is probably innate and that we share with infants and other animals. Such a capacity allows us to judge that there are one, two, three, or many items present, and to judge that one array is greater than another. But the capacity to know that $25 + 376$ is *precisely* 401 depends, Dehaene et al. argue, upon the operation of distinct, culturally inculcated, and language-specific abilities.

The evidence for this is threefold. First, there is suggestive data from brain-damaged patients. Some of these patients (with left parietal damage) lose their general sense of number, and cannot decide, for example, whether 9 is closer to 10 or to 5. Yet they can still perform rote arithmetic: they know that 7 × 7 is 49, and so on. Other patients (with left frontal damage) may present the opposite profile and are unable to decide whether 2 + 2 is 3 or 4, but they know that it is closer to 5 than to 9. This already suggests a certain disassociation between capacities for exact and approximate calculation, as if each facility depends on distinct neural resources.

The second source of evidence is experiments on bilinguals. Trained in one language to do specific sums, some requiring approximation and some requiring exact reason, subjects were then tested on the same problems in their other language. For the exact problems, this switch caused increased response times. For the approximate sums, there was no such cost in response time. Again, it looks as if the approximate sums are solved and stored using a language-independent resource, whereas the exact ones depend on some language-specific encoding.

Third, and to my mind most convincing, the researchers used advanced brain-imaging techniques to observe neural activity in volunteers performing a variety of exact and approximate calculations. The exact calculations evoked increased activity in areas of the left frontal lobe known to be speech and language related, whereas the approximate calculations evoked activity in bilateral areas of the parietal lobes, regions known to be important for visuo-spatial reasoning. The kind of mathematical reasoning unique to our species appears to depend, in part, upon *neural representations of number-words*. It depends, therefore, upon a learning cycle that essentially involves experience with one of the most basic and ubiquitous species of cognitive technology: the spoken words of our public language.

Another way to see the importance of this language-specific loop is to reflect on the interesting case of short-term memory (STM) for lists of numbers. At one time, human STM was considered to be bound by "the magic number seven." Based on experiments involving the capacity of subjects to recall presented lists of numbers (e.g., 4, 9, 7, 1, 2, 7, 6, 9, 4) after a brief delay (on the order of twenty seconds), a short-term memory space of about seven items was calculated. Conduct these same experiments in China, however, and the number goes up. Conduct them in a certain Cantonese dialect and it can reach ten or more items! The reason is that such lists are typically recalled as lists of number words, held in a kind of phonetic loop, and Chinese number words are much briefer than English ones. The briefer the sounds, the more number words can be held in the temporary buffer.

Certain aspects of human mathematical intelligence and recall are thus seen to depend on the individual nature of the specific number words themselves.[25]

Advanced mathematics, of course, typically involves not just *experience* with number words but also the production and re-perception of actual, persisting numerical inscriptions: sums and equations written out on paper or blackboard. For most of us, even the task of calculating 796547×4179645 requires the use of some kind of external tool: pen and paper or a calculator perhaps, or both, if you are especially cautious! And it isn't hard to see why. For most of us (idiot savants and highly skilled mathematicians excepted) know, thanks to some early rote learning, that 7×2 is 14. We do not know that 47×42 is 1,974 (Is it? How will you check for yourself?). Hence, when confronted with the larger-scale problem, most of us need to carve it up into bite-size chunks (7×2, then 7×4, and on) according to a specific routine, as taught in school. Moreover, this routine, when the numbers are fairly large, cries out for the use of pen and paper to store various intermediate results. Our biological short-term memory (STM) is not large enough to store and sequence all the intermediate results. Some of us, like the memory-masters at carnivals and fairgrounds, become adept at using clever mnemonic ploys to squeeze items together, thus packing more into biological STM, while others are simply in command of an unusually large store of prelearned mathematical facts and relations. Many professional mathematicians fall into this category, but most of us still resort to pen and paper to solve the larger and more complex problems.

One quite general way to see the contribution of tools such as pen and paper is thus in terms of a deep *complementarity* between what the biological brain is naturally good at, and what the tool provides. Biological brains do not seem to function like logic machines or like digital computers. Brains—unlike computers—are not good at storing and recalling long arbitrary sequences such as a 200-digit number. Brains—again unlike computers—are not good at recalling long arbitrary lists of instructions. That's why a lot of multifunction technology, like current PCs, and a lot of old technology, like first-generation VCRs, can be so hard to use. These technologies require the biological brain to perform a role for which it is inherently unsuited: recalling and executing long, essentially arbitrary lists of instructions. On the other hand, brains, unlike standard digital computers, *are* good at pattern matching and at simple associations (as when the sight of the cat's tail activates your memory of the whole cat, or when the smell of a certain perfume activates a sudden whirl of thoughts and memories). Our brains are also good at perceptual processing, at using sensory input to control bodily movements, at reasoning about location and movement in space, and the like. Our overall profile is indeed "Good at Frisbee, Bad at Logic."[26]

Scaffolded Thinking

For biological brains the question is how to profit from their pattern-associating strengths while minimizing their weaknesses? One excellent strategy is (you guessed it) to combine the biological pattern associating systems with various environmental props, aids, and scaffoldings. Pictures and spoken words, then written words and diagrams, and most recently the full firepower of interchangeable digital media rank high among the tools by which we press maximum problem-solving power from brains like ours.

Seen in this light, one small story told in chapter 1 takes on a new dimension. It was the story of the presentation preparation where, confronted at last with the shiny finished product, the human being (especially if she is a card-carrying physicalist always seeking the basic scientific explanation of everything) may find herself congratulating her *brain* on its good work. But this, we argued, is misleading. It is misleading because the structure, form, and flow of the final product often depends heavily on the complex ways the brain cooperates with, and depends on, various special features of the media and technologies with which it continually interacts. We tend to think of our biological brains as the point source of the whole final content, but if we look a little more closely what we often find is that the biological brain participated in some potent and iterated loops through the cognitive technological environment. These loops can now be seen to consist, in many cases, in the use of the stable external environment as a source of *complementary* capacities to those provided by the biological brain. We began, perhaps, by looking over some old notes, then turned to some original sources. As we read, our brain generated a few fragmentary, on-the-spot responses, which were duly stored as marks on the page or in the margins. This cycle then repeats, pausing to loop back to the original plans and sketches, amending them in the same fragmentary, on-the-spot fashion. The whole process of critiquing, rearranging, streamlining, and linking is deeply informed by specific properties of the external media, which allow the sequence of simple, pattern-associative reactions to become steadily organized and to grow (hopefully) into something like an argument or presentation. The brain's role is crucial and special, but it is not the whole story. In fact, the true power and beauty of the brain's role was that it acted as a mediating factor in a wide variety of complex and iterated processes, which continually looped between brain, body, and technological environment, and it is this larger system that solved the problem.

Consider now a superficially very different kind of case, the role of sketching in certain processes of artistic creation. Van Leeuwen, Verstijnen, and Hekkert offer a careful account of the creation of certain forms of abstract art, depicting such creation as heavily dependent upon "an interactive process of imagining, sketching and evaluating [then re-sketching, re-evaluating, etc.]"[27] The question the authors pursue is, Why the need to sketch? Why not simply imagine the final artwork "in the mind's eye" and then execute it directly on the canvas? The answer they develop, in great detail and using multiple real case studies, is that human thought is constrained, in mental imagery, in some very specific ways in which it is *not* constrained during online perception. In particular, our mental images seem to be more interpretatively fixed, and less able to reveal novel forms and components. Suggestive evidence for such constraints includes the intriguing demonstration that it is much harder to discover (for the first time) the second interpretation of an ambiguous figure (such as the duck/rabbit in fig 3.3) in recall and imagination than when confronted with a real drawing.[28] Good imagers, who proved unable to discover a second interpretation in the mind's eye, were able nonetheless to first draw what they had seen from memory and then, by perceptually inspecting their own memory-based drawing, find the second interpretation!

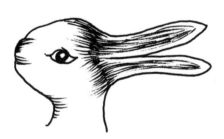

Fig. 3.3   The Duck/Rabbit.
Artwork courtesy of Christine Clark.

Certain forms of abstract art, Van Leeuwen et al. go on to argue, likewise depend heavily on the deliberate creation of "multilayered meanings"—

cases where a visual form, on continued inspection, supports multiple different structural interpretations.[29] Given the evident constraints on our ability to find new interpretations using mental imagery alone, it is not surprising that the discovery of such multiple interpretable forms turns out to depend heavily on a kind of looping process. In this looping process the artist first sketches and then perceptually, not merely imaginatively, re-encounters visual forms, which she can then inspect, tweak, and re-sketch so as to create a final product that supports a densely multilayered set of structural interpretations. The fossil trail of this process remains visible in the sequence of sketches themselves. This description of artistic creativity is strikingly similar, it seems to me, to the story about the presentation. The sketch pad is not just a convenience for the artist, nor simply a kind of external memory or durable medium for the storage of fully formed ideas. Instead, the iterated process of externalizing and re-perceiving turns out to be integral to the process of artistic cognition itself.

To dramatize the point, imagine that we encounter a colony of Martian artists. The brains of these artists are very much like ours, but by some freak of evolution, the Martians also possess a kind of biological scratch-pad memory, which allows them to do in their heads what we do using the sketch pad. We would have no hesitation in treating this internal resource as an aspect of the Martian mind. But why, then (aside from the prejudice that all real thinking and cognition must go on inside the ancient biological skin-bag) should we not treat the human artist, armed with her trusty sketch pad, as a unified, extended cognitive system in just the same way? We must never underestimate the extent to which our own abilities as artists, poets, mathematicians, and the like can be informed by our use of external props and media. Such "mind-tools" (I borrow the term from Daniel Dennett, who in turn cites the work of psychologist Richard Gregory) effectively transform complex problems into ones that the biological brain is better equipped to solve. In the words of cognitive anthropologist Ed Hutchins such tools

> permit the [users] to do the tasks that need to be done while doing the kinds of things people are good at: recognizing patterns, modeling simple dynamics of the world, and manipulating objects in the environment.[30]

Good, potentially transparent cognitive tools of all types display this kind of profile. Using such tools requires the biological brain to do only what would come relatively naturally. Yet the tool itself provides means of encoding, storing, manipulating, and transforming data that the biological brain would find hard, time consuming, or even impossible.

In sum, one large jump or discontinuity in human cognitive evolution seems to involve the distinctive way human brains repeatedly create and exploit various species of cognitive technology so as to expand and reshape the space of human reason. We—more than any other creature on the planet—deploy nonbiological elements (instruments, media, notations) to *complement* our basic biological modes of processing, creating extended cognitive systems whose computational and problem-solving profiles are quite different from those of the naked brain.

Our discussion of human mathematical competence displays this process in a kind of microcosm. Our distinctive mathematical prowess depends on a complex web of biological, cultural, and technological contributions. First, the biological brain commands an approximate sense of simple numerosity. Second, specific cultures have coined and passed on specific number words and labels, including key innovations such as words for zero and infinity. Third, the cultural practice of enforcing simple rote-learning regimes (mathematical tables and so forth) added another element to the matrix. Finally, mix in the novel resource of pen and paper, and PRESTO! Our culturally enhanced biological brains can begin to tackle and solve ever-more-complex problems, eventually scaling mathematical heights that unaided biological brains (of our stripe) could never have hoped to conquer.

In all this we discern two distinct, but deeply interanimated, ways in which biological cognition leans on cultural and environmental structures. One way involves a *developmental* loop, in which exposure to external symbols adds something to the brain's own inner toolkit. The other involves a *persisting* loop, in which ongoing neural activity becomes geared to the presence of specific external tools and media.

The deepest contribution of speech and language to human thought, however, may be something so large and fundamental that it is sometimes hard to see it at all! For it is our linguistic capacities, I have long suspected, that allow us to think and reason about our own thinking and reasoning. And it is this capacity, in turn, that may have been the crucial foot-in-the-door for the culturally transmitted process of designer-environment construction: the process of deliberately building better worlds to think in.

How so? The reason is straightforward. When we freeze a thought or idea in words, we create a new object upon which to direct our critical attention. Instead of just having thoughts about the world, we can then make those very thoughts (and thought processes) the targets of more thinking. This opens up the space of what I call "second-order cognitive dynamics." By this I mean that it opens up the possibility of thinking about how to think well, and allows us explicitly to ask things like this:

What is my reason for believing that?
Is it a good reason?
How sound is the evidence?
How could I gather better evidence?
Under what circumstances do I think best, and how can I bring them about?
How can I build a better world in which to think and reason?

The list could be continued but the pattern is clear. In all these cases we are effectively thinking about our own cognitive profile or about specific thoughts.[31] Second-order cognitive dynamics, I suggest, are possible only once a resource such as language allows us to make our own thought processes into objects for further scrutiny—only when, as Daniel Dennett puts it, we command "a representation of the reason [which may be] composed, designed, edited, revised, manipulated, endorsed."[32]

Donald Merlin, in his excellent exploratory text *The Making of the Modern Mind*, usefully distinguishes two ways of using speech and language. They are the mythic, and the theoretic. Mythic uses focus on storytelling and narrative. The Greeks, however, are said to have begun the process of using the written word for a new and more transformative purpose. They began to use writing to record ongoing processes of thought and theory-building. Instead of just recording and passing on whole theories and cosmologies, text began to be used to record half-finished arguments and as a means of soliciting new evidence for and against emerging ideas. Ideas could then be refined, completed, or rejected by the work of many hands separated in space and time. What was thus created, Donald argues, was

much more than a symbolic invention, like the alphabet or a specific external memory medium, such as improved paper or printing. [It was] the process of externally encoded cognitive change and discovery.[33]

THE EARLY ADOPTER'S DREAM TECHNOLOGY

It was hard to believe. A fully portable, shareable resource, which would radically alter the way we think, work, and live. The early adopters, indeed, would be so vastly empowered that there were great fears in the land concerning fairness, access, and equality. Subject to local protocol matches, groups of users could cheaply share information and coordinate activities across vast disconnections in space and time. Totally human-centered, delicately matched to the strengths and weaknesses of our biological brains, able to evolve and alter to become easier to learn and deploy, the new piece of kit was, in fact, so simple that even a child could use it! Yet it would allow us to learn quicker, to grasp concepts otherwise beyond our reach. And—wonder of wonders—it would allow us to begin actively to think about our own thoughts and problem-solving strategies. As a result, it would invite us to systematically and repeatedly build *better worlds to think in.*

Many feared the new resource. They felt it was sure to encourage great laziness and to stop people thinking for themselves. If you could just ask someone for the answer, who would bother to learn anything? In the presence of such potent resources, wouldn't our "real" memories simply wither away? Where would it all lead? Might we not turn into a race of lazy, desensitized "post-humans"—hybrids who had traded flesh and spirit for artifice, abstraction, and power?

You be the judge. For the technology was (of course) language, and indeed, it changed us beyond recognition. It brought into being the kinds of explicit thought and reflection upon which this whole scenario depends. That's why the scenes just imagined could never have occurred. Public language was the spark that lit the hybrid fire.

(Facing page)   Fig. 3.4   The mangrove tree builds islands by catching floating debris in long vertical roots shot through the water from floating seeds. The tree thus builds the ground it seems to stand on. Could words, by a similar trick, sometimes build thoughts rather than merely express them? Artwork courtesy of Christine Clark.

The process of which Donald speaks is the public, collective version of the kind of scaffolded thinking and reasoning described earlier. Just as I might use pen and paper to freeze my own half-baked thoughts, turning them into stable objects for further thought and reflection, so we (as a society) learned to use the written word to power a process of collective thinking and critical reason. The tools of text (and to some extent speech) thus allow us, at multiple scales, to create new stable objects for critical activity.

With speech, text, and the tradition of using them as critical tools under our belts, humankind entered the first phase of its cyborg existence. What we had succeeded in doing was to discover and harness a new kind of cognitive resource: a kind of magic trick by which to go beyond the bounds of our animal natures. One image that I find useful in thinking about this is the image of the Mangrove Swamp.

Picture yourself in the humid swamplands of the area known as Ten Thousand Islands—a maze of black mangroves extending from Key West to the Everglades. You are stunned by the distribution and density of these unusual trees, some of which reach heights of more than eighty feet. Yet often, these large trees stand neatly, one per island, on their own small beds of land. How did this neat arrangement arise? The answer is unexpected, for the trees did not seed upon the islands. Instead, the islands were built by the trees.[34] The mangrove (fig 3.4) grows from a floating seed

that sends complex vertical roots through the water, searching for shallow mud flats. The first result looks like a tree on stilts in the water, a bit like those famous swamp houses seen in many a Hollywood movie. But quite soon, the raised roots collect dirt and debris carried through the water, and a small island begins to form. Sometimes several such islands merge creating a new shoreline. In these swamplands, our standard expectations (that trees need land to grow on) are upset. Most of the visible land is built by the trees. The tree comes first, the island second.

In much the same way, I suggest, we tend to think of words and language as simply built upon the preexisting islands of our intelligence and thought. But sometimes, perhaps, the cycle of influence runs the other way. Our words and inscriptions are the floating roots that actively capture the cognitive debris from which we build new thoughts and ideas. Instead of seeing our words and texts as simply the outward manifestations of our biological reason, we may find whole edifices of thought and reason accreting only courtesy of the stable structures provided by words and texts. We saw something of this process in the chimps' use of the concrete labels for sameness and difference. We see something of this in the manager's construction of a presentation, the artist's use of a sketch pad, and the mathematician's use of number words and external encodings. Over time we become sensitized to the relation between good cognitive products and the processes that gave rise to them. We then begin to actively structure our worlds (from our schools, to our offices, to our peer review systems) in ways that help promote better thinking. Soon, we inhabit a world not simply adapted to our bodily needs (with heating, clothes, and cooking) but to our cognitive strengths and weaknesses. All of art, science, education, and culture, I shamelessly speculate, is testimony to this runaway process. Human cognition is now a moving target. The biological organism is just one part of the chameleon circuitry of thought and reason, much of which now runs and flows outside the head and through our social, technological, and cultural scaffoldings.

Laying all this at the foot of the door of language may seem to many to go too far. We are already pretty special, after all, in being able to use and understand human language at all. Maybe it isn't the language that makes us smart so much as the smartness that lets us learn and use language. As so often, the truth surely lies in between. Something, clearly, allows us to learn the kind of open-ended structured language that sets us apart from most other animals, but it is only because we command such a distinctive resource that we become able to treat our own thoughts and ideas as objects. It is this process—of using words to turn thoughts and ideas into new stable objects for further thinking and reasoning—that starts the real cognitive snowball rolling.

Building Better Brains

To all this, we now add a final neurobiological ingredient. There is growing evidence that the human brain, more than that of any other animal on the planet, benefits from what has become known as constructive learning. A constructive learning system, broadly speaking, is one whose own basic computational and representational resources alter and expand (or contract) as the system learns. To get the idea, consider the contrast between a system with a fixed short-term memory (STM), and a system in which the capacity of the STM gradually increases over some developmental period. Or a system with fixed speed and processing power, versus one in which speed and processing power can be increased. Or a system with a fixed stock of representational resources (like a fixed dictionary of words), versus one capable of adding brand new items to its dictionary/vocabulary as required.

Constructive learning systems use early learning to build new basic structures upon which to base later learning. There is a powerful body of computational work using artificial neural networks, which has begun to show in concrete detail how such increases in problem-solving capacity can be systematically achieved. There is work that shows, for example, that an artificial neural network whose STM grows as it learns can solve problems that defeat a fixed-architecture learner.[35] There is similar work using networks, which add basic processing units and connections during learning.[36] What matters most in all these scenarios is the system's capacity to build an internal representational and computational environment that is *itself* a partial response to the early training environment. The structure of the encountered problem domain thus determines, to some extent, the architecture (number of units, layers, connections) of the network.

Might anything remotely similar go on inside the human brain? Neural constructivists, such as Steve Quartz and Terry Sejnowski, think the answer is yes. More precisely, their claim is that similar mechanisms of neural growth allow the human cortex to function as an "organ of plasticity," which is shaped and sculpted by the problems, resources, and opportunities encountered during postnatal and lifetime learning.[37] This means that the environments in which our brains grow and develop may actually help structure the brain in quite deep and profound ways.

The evolutionary emergence of the mammalian neocortex is generally accepted as the key neural innovation underlying advanced reason. Cortical evolution, if the neural constructivists are correct, is not simply a story about the addition of new, special-purpose brain structures. Rather, it is a story about the addition of a plastic resource geared to allowing the encountered environment to build dedicated, delicately fitted neural substructures "on-the-hoof." The human neocortex and prefrontal cortex, along with the extended developmental period of human childhood, allows the *contemporary* environment an opportunity to partially redesign aspects of our basic neural hardware itself. The designer environments discussed in the previous chapters are thus matched, step-by-step, by dedicated designer brains, with each side of the co-adaptive equation growing, changing, and evolving to better fit—and maximally exploit—the other. It is in this way that the human learner becomes "dovetailed" to the set of reliable external problem-solving resources that she encounters during early learning.

The neural constructivist vision thus depicts neural and especially cortical growth as experience-dependent and involving the actual construction of new neural circuitry (synapses, axons, dendrites) rather than just the fine-tuning of circuitry whose basic shape and function is already determined. The learning device *itself* changes as a result of organism-environmental interactions; learning does not just alter the knowledge base for a fixed computational engine, it alters the internal computational architecture itself.[38]

As a concrete example, consider the development of hearing. Congenitally deaf children, whose brains are thus never exposed to the complex and distinctly structured inputs that the auditory world provides, fail to develop the complex web of inner connectivity that supports normal hearing. If such stimulation is artificially provided, using the kind of cochlear implant described in chapter 1, recovery is rapid. The neural bases of this recovery are increasingly well understood and involve complex changes in the connectivity and response characteristics of auditory cortex. Visual cortex, likewise, requires extensive, experience-dependent rewiring to support seeing. Newborn human infants have very bad vision; it is highly restricted in scope, and the resolution is forty times worse than adult vision. Depth appreciation is pretty well nonexistent. It takes about a year of "cortical training" for the visual system to become normal, a process that can be blocked by cataracts or other impairments, which deprive the visual cortex of the experience it needs. Remove the cataracts and replace the affected lens with a clear artificial one, and improvement is again dramatically fast. According to one researcher, this kind of result "demonstrates the amazing plasticity of the young brain and underscores the importance of complex, balanced, early sensory input for guiding subsequent brain development."[39]

So great, in fact, is the plasticity of immature cortex (and especially that of prefrontal cortex, according to Quartz and Sejnowski) that O'Leary dubs it "protocortex." The whole sensory, linguistic, and technological environment in which the human brain grows and develops is thus poised to function as one of the anchor points around which such flexible neural resources adapt and fit. Such neural plasticity is, of course, not restricted to the human species; in fact, some of the early work on cortical transplants was performed on rats. But our brains do appear to be far and away the most plastic of them all. Combined with this plasticity, however, we benefit from a unique kind of developmental space: the unusually protracted human childhood.

In a recent evolutionary account, Griffiths and Stotz argue that the long human childhood provides a unique window of opportunity in which "cultural scaffolding [can] change the dynamics of the cognitive system in a way that opens up new cognitive possibilities." These authors argue against what they describe as the "dualist account of human biology and human culture" according to which biological evolution must first create the "anatomically modern human" before being followed by the long and ongoing process of cultural evolution. Such a picture, they suggest, invites us to believe in something like a basic biological human nature, gradually co-opted and obscured by the trappings and effects of culture and society. This vision (which is perhaps not so far removed from that found in some of the more excessive versions of evolutionary psychology) is akin, they argue, to looking for the true nature of the ant by "removing the distorting influence of the nest."[40]

A more realistic vision depicts us humans as, by nature, products of a complex and heterogeneous developmental matrix in which culture, technology, and biology are pretty well inextricably intermingled. It is a mistake to posit a biologically fixed "human nature" with a simple wrap-around of tools and culture; the tools and culture are indeed as much determiners of our nature as products of it. Ours are (by nature) unusually plastic and opportunistic brains whose biological proper functioning has always involved the recruitment and exploitation of nonbiological props and scaffolds.[41]

From this neurologically and ecologically unique whirlpool, we humans emerge. We are beings factory-tweaked and primed in order to be ready to participate in hybrid cognitive and computational regimes, able to think and learn in ways that take us, bit-by-bit, far beyond the scope and limits of our basic biological endowments. To be sure, there is a basic profile of biological strengths and weaknesses, one that, as Hutchins and Norman both suggested, must act as a kind of reference point for our technologies. We should not underestimate the capacity of human brains in general—young human brains in particular—to simultaneously alter and grow so they can better exploit the problem-solving opportunities our technologies provide.

We see this in the physical domain every day. A recent Warwick University study showed that young people's thumbs have overtaken fingers as the most muscled and dextrous digits among the under-twenty-fives, simply as a result of their extensive use of handheld electronic game controllers and text messaging on cell phones. New generations of phones will be designed around this greater agility, leading to even further changes in manual dexterity and the like, in a golden loop. The same kind of user-technology co-adaptation can occur at the deepest levels of neural processing. Such developmentally open brains are not just opportunistic, but *explosively* opportunistic. They are ready to change themselves to make the most of the structures, media, and opportunities encountered during learning.

Such explosive opportunism has implications for social policy and educational practice. The goal of early education (and perhaps of *all* education) should not be seen as simply that of training brains whose basic potential is already determined. Rather, the goal is to provide rich environments in which to *grow* better brains. The more seriously we take the notion of the brain-environment *engagement* as crucial, the less sense it makes to wonder about the relative *size* of each of the two contributions. What really matters is the complex reciprocal dance in which the brain tailors its activity to a technological and sociocultural environment, which—in concert with other brains—it simultaneously alters and amends.[42] Human intelligence owes just about everything to this looping process of mutual accommodation.

Our brief foray inward is now at an end. We have glimpsed just a few of the biological innovations that help make us so culturally and technologically open. One is our capacity to re-create our own body image on the hoof. This capacity is especially important in allowing us to imaginatively relocate ourselves courtesy of new techniques such as telepresence and telerobotics. Our brain is highly opportunistic, ready and willing to allow reliable external structures to do duty both as memory store and as processing arena. The worlds of speech and text here play a special role. Sometimes we internalize strategies that originally involved the actual manipulation of external symbols and objects. At other times we learn strategies that will require the continued presence of various external scaffoldings and support (pens, paper, and so forth). While all this goes on, if the neural constructivists are correct, we remain open to quite profound kinds of neural (cortical) growth and rewiring. In all these ways we are transformed by the almost unimaginable effects of our own primary transition technologies. The biggest transformation of all, however, was the one that occurred when our thoughts and ideas became objects of our own critical attention. By making our own thoughts into stable objects for our own and others' unhurried scrutiny, our skills with language opened the floodgates of self-reflective reason. We began to think about our own thoughts and about how to build better tools for thinking. Revamped and enhanced, all bets were off. Human cognition was poised to go indefinitely beyond its animal origins.

CHAPTER 4

Where Are We?

My body is wherever there is something to be done.

—Maurice Merleau-Ponty

Stretch to Fit

"Distance," a philosopher-friend once commented, "is what there is no action at."[1] There is considerable wisdom in this. Next time you are on a crowded train or in a subway station, look at all the people around you talking on their cell phones. Where are they? Well, clearly, they are with you in the station or on the train, but often, they are not much engaged with these local surroundings. They are, temporarily at least, jacked into a web of personal and business communications, which deliberately disrespects current physical location. Draw the lines of proximity and distance according to the criterion of effective action, and a virtual neighborhood emerges; one in which the speakers are more proximal to their colleagues or loved ones than to the strangers on the platform.

There is nothing especially new or surprising in this. Our sense of our own location, like our sense of our own bodily limits as discussed in chapter 3, is the fruit of an ongoing project. It too is a construct: this time, one formed by our implicit awareness of our current set of potentials for action, social engagement, and intervention. Imagine yourself confined to a hospital bed. You cannot walk, but you can move your arms and hands. Your world seems to shrink to the radius of action and control. Add a buzzer to summon a nurse and you feel a tad more liberated. Add a phone link to your stockbroker and/or your family, and the claustrophobia recedes even more. But action, of the kind that seems most important for our sense of our own location, is a complex thing. The mere provision of telecommunication links, though it goes some way toward freeing us from the bonds of physical space and proximity, is not really enough to alter our bedrock sense of where we are. In this chapter, I want to explore the potential of richer technologies to impact, for better and worse, this fundamental sense of location. In the next chapter, I explore the potential of these technologies to alter our fundamental sense of self.

In my opinion, the single best piece of philosophical fiction ever written must be the short story "Where Am I," which Daniel Dennett (professor of philosophy at Tufts University) published in 1981. Dennett tells the story of an American citizen who agrees to participate in a secret experiment. The citizen is Dennett himself, and in the experiment Dennett's brain is removed, kept alive in a tank of nutrients, and equipped with a multitude of radio links by means of which to execute all its normal bodily control functions. Dennett's body is equipped with receivers and transmitters, so that it can use its built-in sensors (eyes, ears, etc.) to relay information back to Dennett's brain. As the technicians in the story put it:

> Think of it . . . as a mere *stretching* of the nerves. If your brain were just moved over an *inch* in your skull, that would not alter or impair your mind. We're simply going to make the nerves indefinitely elastic by splicing radio links into them.[2]

His brain safely excised and relocated, and the radio links established, Dennett awakes. He sees the nurse, who leads him to the room where his brain is being kept. The experience that ensues is puzzling. There is Dennett, standing up, staring at his own brain. Or is he? Perhaps, he muses, the proper thought is that "here I am, suspended in a bubbly fluid, being stared at by my own eyes." But try as he may, Dennett cannot seem to place himself *in the tank*. It continues to seem as if he is outside the tank, looking in. Dennett's point of view, as he moves, seems securely fixed outside the tank. The feeling shifts, however, when Dennett's body is subsequently trapped by a rockslide, entombed far beneath the earth's surface. At first, Dennett feels trapped beneath the surface, but then the radio links themselves begin to give way, rendering him blind, deaf, and incapable of feeling. The shift in point of view was immediate.

> Whereas an instant before I had been buried alive in Oklahoma, now I was disembodied in Houston. . . . as the last radio signal between Tulsa and Houston died away, had I not changed location from Tulsa to Houston at the speed of light?[3]

Where would you place our hero? Is Dennett *really* in the tank of nutrients, *really* trapped beneath the soil, or really no place at all (or both places at once)? Such questions need have no clear-cut answers. But what does seem clear is that our sense of location is not simply a function of our beliefs about the location of our body. Dennett, after all, continues to believe that his body is buried in Oklahoma, but his *point of view* is more labile. It is, I want to say, a construct grounded in the brain's experiences of *control, communication, and feedback*. And as such, it is open to rapid and radical reconfiguration by new technologies.

Dennett's story was pure fiction, but science is never far behind. Consider the work of Miguel Nicolelis of Duke University in North Carolina.[4] Nicolelis and his team studied the way signals from the cerebral cortex control the motions of a monkey's limbs. An owl monkey had ninety-six wires implanted into its frontal cortex, feeding signals into a computer. As the monkey's brain sent signals to move the monkey's limbs, this "neural wiretap" was used to gather data about the correlations between patterns of neural signal and specific motions. The correlations were not simple and turned out to involve ensembles of neurons in multiple cortical areas, but the patterns, though buried, were there in the signals. Once these mappings were known, the computer could predict the intended movements directly from the neural activity. The computer could then use the neural signal to specify movements, which could be carried out by a robot arm. In experiments conducted with the MIT Touch Lab, signals from the owl monkey's brain in North Carolina were used to directly control an electromechanical prosthesis in an MIT laboratory six hundred miles distant.

The results were impressive. The neural commands were rapidly and accurately translated into actual motions of the remote robot arm, which mimicked the full range of motions of its biological template. Dr. Mandayam Srinivasan, director of the Touch Lab,[5] commented that "it was an amazing sight to see the robot in our lab move, knowing it was being driven by signals from a monkey brain at Duke. It was as if the monkey had a 600 mile-long virtual arm."[6] The robot thus controlled was a haptic interface, part of a multisensory virtual reality system used to touch, feel, and manipulate computer-generated objects. Pursuing the theme, Dr. Srinivasan speculates that "if we extended the capabilities of the arm by engineering different types of feedback to the monkey—such as visual images, auditory stimuli and forces associated with feeling textures and manipulating objects—such closed-loop control might result in the remote arm's being incorporated into the body's representation in the brain." In short, there may be all kinds of ways in which we can one day augment our bodies in virtual space, extending and altering our own body image and representation into the bargain.

There is, of course, a whole swathe of technologies supporting so-called telepresence, literally, remote presence. The good old-fashioned telephone affords a thin, narrow bandwidth kind of aural telepresence. Typically, the term implies state-of-the-art equipment supporting a more realistic, multidimensional effect. Taken to the limit, the effect would be very much as in Dennett's thought experiment, except that instead of going to all the trouble of removing the brain and setting it up to control and communicate with the distant body, the technologies of telepresence leave brain and body joined and intact but wrap the body in a kind of additional sensory cocoon.[7] This cocoon is fed with information gathered—perhaps using a mobile robot body—at some distant site. That information is used to power (via the cocoon) a local sensory barrage corresponding to the distally detected inputs.

The term "telepresence" was introduced into the literature in 1980 by the computer scientist and A.I. (Artificial Intelligence) pioneer Marvin Minsky.[8] Minsky's inspiration was the kind of tele-operation system in which a worker might handle radioactive materials by wearing a pair of special lenses, along with gloves and sleeves that transmit her arm and hand motions to a robotic device. The device in turn transmits visual and tactile feedback to the operator. In this way, it is as if the operator were actually present in the hazardous environment. In such cases operators report that they rapidly and effectively come to feel the shift in point of view so vividly described by Dennett, flipping back and forth between the local and the distant locations as needed.

True telepresence, insofar as it is achievable, would seem to require a high bandwidth multisensory bath of information with local sensory stimulation: in effect, the full virtual reality bodysuit, with feedout and feedback connections for sight, sound, hearing, touch, and smell, as well as heat and resistance sensing. Also—perhaps crucially—the user needs the ability not just to passively perceive but to *act upon* the distant environment and to command the distant sensors to scan intelligently around the scene.

It is noteworthy, however, that temporary shifts in point of view can be achieved using much more limited and ordinary resources. If you visit the Virtual Artists' VA Robocam Site (http://www.robocam.va.con.au/) you can interact with a motorized camera mounted on a tall building, sweeping the area as you desire.[9] A similar project was described by Minsky in the original *Omni* paper like this:

> A Philco engineer named Steve Moulton made a nice telepresence eye. He mounted a TV camera atop a building and wore a helmet so that when he moved his head, the camera on top of the building moved, and so did a viewing screen attached to the helmet. Wearing this helmet you have the feeling of being on top of the building and looking around Philadelphia. If you "lean over" it's kind of creepy. But the most sensational thing Moulton did was to put a two-to-one ratio on the neck, so that when you turn your head 30 degrees, the mounted eye turns 60 degrees: you feel as if you had a rubber neck, as if you could turn your "head" completely around.[10]

This description highlights two important points. The first is that even quite basic *but interactive* technologies can generate a sense of real telepresence. Comparing the VA robocam experience (visit the telepresence hub at http://mitpress.mit.edu/telepistemology) with an experience of purely passive viewing (e.g., the wonderful web camera that looks at the African landscape: www.africam.com) is instructive. The passive experience leaves the observer clearly at home; it is no more like telepresence than looking at the photos in *National Geographic* (though it *is* sometimes more exciting).

Yet as soon as a distant camera responds to your controls, and especially if the mode of control is either natural (the helmet rig) or highly practical (a gamester with a joystick), you begin to feel relocated, as if you are in the distant scene.

Given our discussion in chapter 3, this should come as no surprise. There we saw that normal human vision involves a complex process of intelligent search and just-in-time information retrieval. In normal vision we leave most of the information out in the world, secure that we can, with a flick of the eye, retrieve what we need to know as and when required. When faced with input from a fixed camera much of that flexibility is lost. As a result, the scene presents itself as a source of visual information, but not really as a context for fluent, embodied action. Our sense of personal location has more to do with this sense of an *action-space* than with anything else.

The links between our capacities for action and our perceptual experiences are extraordinarily deep and potent. In a famous series of psychological experiments, human subjects were fitted with special glasses whose lenses turned the visual input upside down. At first, as you would expect, the subjects saw an upside-down world, but after a period of sustained use, the visual world began to "flip back over." After a few days, the subjects reported that their visual experience was back to normal. Remove the glasses, however, and the world now looks upside down (for a while, until re-adaptation occurs). Most interesting of all, these kinds of perceptual adaptations are highly action-dependent. They are primarily driven by the *combination* of the visual inputs and the subject's experiences of trying to move and act in the world (and hence, crucially, by feedback coming through various motor and locomotion systems). Thus a subject fitted with the lenses, but simply pushed around in a wheelchair, does not show the adaptation, while one who walks along a complex trail does.[11] So fluent are our perceptual systems at making these motor-loop-dependent adaptations that it is even possible to adapt to both the presence and the absence of the lenses. By wearing the goggles intermittently, while acting in the world, you can train your visual system to cope with both kinds of input (right way up and upside down). This coping is, remarkably, quite seamless. The instant you don or remove the goggles, your visual system flips into one of the two "settings." The scene before your eyes looks unchanged to you, nothing seems to flip or alter; ask an untrained friend to try it, and she will immediately flounder in the face of the upside-down world![12]

Interestingly, such adaptation need not be global. Instead, there can be adaptation for certain well-practiced motor routines and not for others. Subjects fitted with *sideways* shifting lenses (those that shift the image a little way to the left or right) who played repeated games of darts displayed adaptation *only* while using their normal dart throw. If they were then asked to throw underhand, or with their left hand (if they were right-handed), the compensatory effects of the practice immediately disappeared.[13] These results further underline Ramachandran's principle, as described in chapter 3. The principle, remember, was that the "mechanisms of perception are mainly involved in extracting statistical correlations from the world to create a model that is temporarily useful."[14] The most important of these correlations—as the nose-tapping experiments already suggested—are those between perceptual inputs and our own deliberate motions and actions.

The notion that our perceptual experience is determined by the passive receipt of information, though seductive, is deeply misleading. Our brains are not at all like radio or television receivers, which simply take incoming signals and turn them into some kind of visual or auditory display. Who would there be to look at the display anyway? The whole business of seeing and perceiving our world is bound up with the business of acting upon, and intervening in, our worlds. And where action and intervention goes, our sense of bodily presence and location swiftly follows. The extent to which current efforts at telepresence support appropriate kinds of fluent action and intervention is, however, rather limited. Here are some fairly representative examples:[15]

(1) Sandpit Excavation
The first documented internet-based telebot was set up in 1994 at the University of Southern California. The remote user could control a digital camera and airjet, mounted on a robot arm so as to "dig for artifacts" in a sandbox in the USC lab.[16]

(2) Bird Brains
In 1996, Eduardo Kac, an artist, writer, and media theorist, set up an "interactive networked telepresence" installation at the Nexus Contemporary Art Center in Atlanta. There, you saw a large (real) aviary stocked with thirty

flying birds and one large robot bird. In front of the aviary was a virtual reality headset. Wearing the headset, the viewer was able to perceive the aviary from within the robot bird. The bird's eyes were twin digital cameras, and the viewer's head movements moved the head of the bird. The viewers could even watch themselves, standing outside the cage and wearing the headgear, in this way! The upshot, according to Kac, was that "the local participant [was] both vicariously inside and physically outside the cage . . . a metaphor that revealed how new communications technology enables the effacement of boundaries at the same time as it reaffirms them [and addresses] issues of identity and alterity."[17]

(3) GARDENER'S WORLDS

The telegarden (Goldberg et al., 1994) is a telerobotic, web-accessed, yet totally real garden. Visitors use CCD cameras and a robot arm to plant seeds or plants, water them, pull them up, and so forth. The idea is to "invite participation" and encourage return visits and monitoring. You can read about the garden at http://telegarden.aec.at.

(4) TELEBOTIC TILLIE

Go to www.lynnhershman.com/tillie and visit Lynn Hershman's San Francisco gallery through the eyes of Tillie, a telerobotic female doll (see fig 4.1). Click on her eye and you will see what is currently visible to the camera in Tillie's eye as she sits in the gallery. You can turn her head, look around, and so on. You can also view the gallery "objectively," seeing Tillie herself watching you, perhaps with your own eyes!

The world of industrial telerobotics is, not surprisingly, somewhat more advanced. By "industrial robotics" I mean both what is more properly termed "teleoperation" and genuine "telerobotics."[18] Teleoperator systems are ones in which the human "master" directly guides a distant robotic "slave." The slave is meant to merely echo, at a distance, the actions actually being performed by the human master. Early versions of such teleoperator systems (called "telemanipulators") would have been standard fodder at Los Alamos during the time of the Manhattan project, and some, for all I know, might even have found their way into Ed Groshus's Black Hole (mentioned in chapter 2). They were originally developed to allow workers to manipulate toxic or radioactive materials, and they represented a modest advance

Tillie will be at the
Austin Museum of Art!

Fig. 4.1 Lynn Hershman's internet installation "Tillie the Telerobotic Doll." Shown here advertising her presence in the Virtual Embrace exhibition at the Austin Museum of Art, Austin, Texas, in September 2001. By clicking on one of her "eye" cons, web-based viewers could see what she sees and move her head left and right. Photo courtesy of Lynn Hershman.

on the use of simple tong-like appendages. By 1950 there were quite advanced mechanical systems, which translated the movements of an operator on one side of a one-meter-thick quartz window into subtle and precise movements of an identical mechanism on the other side. The first true teleoperation system, however, was invented by Ray Goertz in 1952.[19] Unlike the simple telemanipulators, these systems incorporated advanced electronics and computer control; like the earlier systems, they involved a master and a slave manipulator. Motors, sensors, and calculating devices were also brought into play. Slave-side motors allowed the system to apply significant forces over much larger distances, while master-side motors supported force feedback so that the operator could feel the resistance and compliance of the distant materials.

The 1970s saw the widespread introduction of coordinate transforming teleoperators. These were systems in which the master motions were not simply replicated but were instead systematically transformed (scaled up, scaled down, made to fit within a confined space, etc.). Now, for the first time, the kinematics of the master and slave could diverge.

By the 1980s, true bidirectional teleoperators existed, with the full six degrees of freedom (six independent motion axes) required for the fluent manipulation of rigid 3D objects. The first such device was designed by Bejczy and Salisbury in 1983.[20] It operated on the standard master and slave principle. The human operator places her hand in the sleeve or exoskeleton that tracks and transmits her movements. The distant slave then reproduces the action. Master-side motors allowed the operator's hand to feel the force exerted on the distant slave.

In the late 1980s and throughout the 1990s, telerobotics came into their own. A telerobotic device is one in which the fine details of action control are left (at least in part, and at times) to the robot itself. The human controller sends only a high-level command, which the robot puts into action. Thus

"telerobotics" technology implies communication on a higher level of abstraction in which the human communicates goals and the slave robot synthesizes a trajectory or plan to meet that goal.[21]

Thus a telerobot on the moon might be told, by a human operator watching a video image in Houston, to acquire the large rock appearing in the

Fig. 4.2 Multipurpose domestic robot designed by Rodney Brooks and iRobot Corps. Described as a "remote presence" robot, the iRobot-LE can climb stairs and traverse most household terrain. The user controls the robot over the web, telling it where to go in the distant house. The top-mounted camera and microphone then send sound and vision back to the user, wherever she is in the world. Photo courtesy of Professor Rodney Brooks.

top right-hand corner of the operator's display. The robot would then locally compute the kind of walk, reach, and grip needed to carry out the task. In the same way, a household robot, controlled over the internet, might be told to go into the living room and transmit pictures of the sleeping cat. The iRobot-LE (fig. 4.2) made by Rodney Brooks's company has carried out similar tasks in a Boston apartment, while under supervisory control from Brooks in Tokyo.

The practical reasons for moving toward telerobotics are obvious. It is easier for the operator to issue only high-level commands, and this may be essential when time-delays are critical and bandwidth limited. As a technology of genuine *telepresence*, however, telerobotics may at first seem less promising than teleoperation. The best teleoperator systems, after all, provide rich capacities of finely directed action and intervention, and a wide spectrum of sensory feedback (e.g., force feedback coordinated with visual feedback). This rich two-way energetic exchange is surely just the kind of link that might allow the distant equipment to become transparent in use, whereas issuing high-level commands, with merely visual feedback, to a distant robot seems less likely to generate any real shift in perspective.

Advanced Telerobotics

Such a diagnosis is, however, still a little too hasty. To see why, we need to revisit our old friend the biological brain. As a first step, however, recall the example (chapter 1) of the car driver who relies on an ABS (Automatic Braking System). Once drivers are accustomed to ABS, they cease to feel as if the braking is in any way "out of their control." Yet the machinery mediating between the action of the foot and the actual braking is now much more intelligent than before, able to adjust and pulse the braking action as required. The presence of such modestly intelligent intermediaries, however, need in no way compromise our sense of direct engagement and control. Such semi-intelligent technologies can become as transparent in use as any others. In fact, we are all intimately familiar with this kind of case, since much of our daily bodily activity (and, indeed, our daily decision making) falls into the same category.

Take the simple (or not-so-simple) act of walking to the store. The last time I walked to the store, the sum total of my conscious, deliberate neural activity amounted to something like this: "Oh dear, we've run out of Guinness. I'll just pop out and pick some up. Hope they've got some in the chiller." The high-level decision thus made, a great deal of subsequent activity was left to the good devices of, well, my good devices. It was left to various neural and biomechanical subsystems operating way beneath the levels of my conscious awareness. I never decided, for example, just how far to swing back my right leg while walking to achieve a steady gait (though I suppose I might have, had we drunk a great deal more Guinness beforehand). I never decided how to move around my head and eyes to spot looming obstacles, or how precisely to time and control the trajectory of my hand and arm while reaching for the beer. In fact, on reflection, most of what I did I seemed to have very little to do with. Even the decision to actually go out and get the beer, although it, at least, was conscious, did not seem to arise *from* any set of previous conscious thoughts. It was just suddenly there, in my head, at the forefront of my thoughts (most decisions, as they say, are born, not made). The conscious self, it quickly appears, is but the tip of the "I" berg; the vast bulk of neural activity leading both to, and away from, this tip is unconscious.

Recent experiments confirm and dramatically extend this general diagnosis. Take a look at the two pairs of center (large) circles displayed in figure 4.3. Which circle strikes you as the largest? In the top case, both center circles are the same size; in the bottom case, the one on the right is larger—but they probably didn't look that way to you. Your conscious perception is misled, it seems, by scaling effects caused by the other smaller circles surrounding the targets. This visual illusion is known as the "Titchener Circles" or Ebbinghaus illusion. In 1995 Aglioti and his colleagues published a suggestive follow-up experiment.[22] In this experiment,

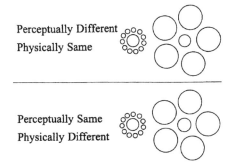

Perceptually Different
Physically Same

Perceptually Same
Physically Different

Fig. 4.3   The Titchener Circles Illusion. In the top figure, the two central disks are the same size, but the one surrounded by the smaller circles looks larger. In the bottom figure, the disk surrounded by large circles is actually larger but now appears equal in size to the other central disk. Figure courtesy of Professor Melvyn Goodale.

thin poker chips were used as the center circles; subjects were asked to physically pick up the chip on the left if the two appeared equal in size, and to pick up the one on the right if they appeared unequal. Subjects used the same hand for each task, and sensitive opto-electronic recording equipment was used to record the precise size of their finger-thumb grip aperture (measured just before actual contact with the disk). As expected, the subjects' choice of chip was influenced by the illusory scaling effect, so they would choose in ways determined by the apparent, not the real, sizes of the disks. But—and here's the punch line—their fine-tuned finger-thumb grip aperture was nonetheless correct. It was quite unaffected by the illusory sizes.[23] The neural subsystems that determined the actual grip aperture were unaffected, it seems, by the high-level illusion about relative size.

The explanation, according to the cognitive neuropsychologists David Milner and Melvin Goodale, is that the human visual system is already a hybrid, a cooperating mixture of two distinct elements. One element is an evolutionarily ancient system for controlling motor actions using visual information. The other, more evolutionarily recent system, takes the same visual inputs but processes them very differently. It extracts information about what the object is (is it a cat, or a cup?), and it makes contact with memory systems (is it an especially heavy cup?) and with reasoning systems (is it covered in oil, and hence slippery?). In extracting this kind of information, Milner and Goodale believe, this system must discard much of the fine detail (precise location in space relative to current arm location, etc.) required to actually *act* on the object. Nature's solution, they argue, is a kind of biological division of labor. One set of neural circuits (the ventral stream), leading from V1 (early vision) to IT (interotemporal cortex), is devoted to recognition, classification, and reasoning. Another set of circuits (the dorsal stream), leading from V1 but proceeding to PP (posterior parietal cortex), is devoted to the fine control of ongoing action (movement). It is this latter system, they claim, which is most directly implicated in conscious seeing and verbal report.[24] The two systems are, however, capable of a kind of limited interaction.

In the case of the Tichener circles it is the conscious (illusory) perception of one circle as larger than the other that causes the autopilot-like subsystem to reach for a specific disk. Our conscious high-level decisions thus serve as the impetus for the other systems to do their stuff, while still devolving substantial subproblems (like the calculation of exact grip and reaching trajectory) to other internal agencies. The conscious self in these cases is exercising a form of what Thomas Sheridan originally dubbed "Supervisory Control": a "type of control in which goals and high level commands are communicated to the slave robot."[25]

Consider the moonrock-gathering telerobot once again. The human operator spots a rock in the top right-hand corner and tells the robot to acquire it. The robot then plans the walk and calibrates the reach. Just how different is this, from the case in which the conscious "I" decides to reach for an object (one of the disks or a can of beer) and nonconscious neural systems kick in to compute arm trajectory, determine grasp size, and so forth? In a recent piece, Mel Goodale suggests that the interplay between the neural systems generating our conscious perceptions and those responsible for the remaining details is thus "reminiscent of the interaction between the human operator and a semi-autonomous robot in what engineers call teleassistance."[26] Ramachandran, the neuroscientist we met in chapter 3, likewise speaks of "the Zombie in the brain," meaning the mass of automatic subsystems, which contribute so profoundly to our thoughts, actions, capacities, and skills. The neuropsychologist Michael Gazzaniga devotes the bulk of his 1998 book The Mind's Past to showing that "even though our sense of purpose and centrality of will are foremost, there dwells within us an automatic and highly specialized machine."[27]

The original cyborg vision, as we saw in chapter 1, was precisely the vision of external, nonbiological elements taking over various automatic functions of the nervous system. At that time, however, attention was largely focused on systems that controlled basic bodily functions such as heart rate and respiration. The goal was to allow the electronically augmented human body to survive in otherwise inhospitable conditions. The full range of tasks that the brain carries out automatically is, however, now known to be much, much larger, and to include many of the operations involved in complex problem solving and even decision making. Knowing this, the range of possible cyborg-like extensions of the human mind expands dramatically. Not just basic physiological homeostasis, but limb control, trajectory planning, and major components of the reasoning process itself may themselves be farmed out. There is no special magic associated with direct physically wired links between components. The differences between links forged by nerves and tendons, by fiber-optic cables, and by radio waves are relevant *only insofar as* they affect the timing, flow, and density of informational exchange. These latter factors are relevant, in turn, because they affect the nature of our relationship with the various kinds of tools, equipment, and subsystems. If the links are sufficiently rich, fluid, bidirectional, fast, and reliable, then the interface between the conscious user and the tool is liable to become transparent, allowing the tool to function more like a proper part of the user. The move thus from teleoperator systems to telerobotics systems relying on high-level commands *need not* result in the alienation of the tool from the conscious user—no more so than the fact that the conscious self merely deciding to go fetch some Guinness results in the alienation of my arms, legs, locomotion, and grasp control systems from the "real me"! In practice, however, teleoperated systems seem to induce the feeling of actual telepresence much more effectively than do existing telerobotic ones. It is time to examine why this is so.

Imagine that you are the human operator of an original Manhattan project telemanipulator. Deep in your B-movie concrete bunker, you handle toxic materials from behind the safety of a thick quartz window. These first-generation devices were clumsy and primitive by today's standards, but despite this "the one-to-one connection between the two sides creates a compelling sensation reproducing the actual sensations of manipulation."[28] The user, in this case, feels as if he or she is actually touching and manipulating the (modestly) distant materials. The blind person whose cane feels like a sensitive extension of her arm is the obvious classic case. You may have had the same experience using chopsticks to select the tastier morsels from the communal platter. Or, when driving your car, you may have had the experience of feeling the road through the system of racks, pinions, axles, and tires.

What seems to matter in these cases is the presence of some kind of local, circular process in which neural commands, motor actions, and sensory feedback are *closely and continuously correlated*. This, of course, is exactly what Ramachandran's principle (which depicts the body image as a temporary construct based on ongoing sensory correlations) would lead us to expect. Remember the compelling demonstrations in chapter 3 where the subjects came to feel as if the desktop or dummy hand were the source of tactile signals being fed to their biological brain?[29] When the dummy or desktop was then hit with a hammer, these subjects showed a galvanic skin response consistent with the expectation of damage to their biological body. They had, at the very deepest level (and after only a few minutes of training), come to identify *themselves* with the nonbiological "extensions." As Ramachandran put it:

> It was as though the table had now become coupled to the students own limbic system and been assimilated into his body image, so much so that pain and threat to the dummy are felt as threats to his own body, as shown by the GSR [Galvanic Skin Response].[30]

I don't think the authors are being entirely facetious when they add that "if this argument is correct, then perhaps it's not all that silly to ask whether you identify with your car. Just punch it to see whether your GSR changes."[31]

In fact, so great is the plasticity of the neural body image that the use of certain augmented reality tools, such as the eyeglass display described in chapter 2, can often cause a temporary distortion of the user's own body image, leading him to make visual mistakes once the eyeglass display is removed.[32] Luckily, sufficient practice under changing conditions (taking the glasses on and off) yields a kind of flippable system able to make the switch without undue cost (rather like those Anglo-American drivers who can switch easily from driving on the right to the left).

In general, then, the sense of extension, alteration, and distal presence arises as a result of close, ongoing correlations between neural commands, motor actions, and (usually multisensory) inputs. Simple telemanipulation and teleoperator systems afford this kind of dense, real-time correlation. The payoff is a compelling sense of bodily augmentation and extension, a sense of genuine, (if modest) telepresence. The intimate web of closely correlated signals and responses necessary for such rarified reinvention of the body is, however, quite fragile and easily disrupted. The most important kind of disruption is temporal: if there is a noticeable time lag between issuing the command and receiving the sensory feedback, or (worse still) if the time lag is variable due to the traffic on phone lines, for instance, the illusion is shattered. This is what happens, then, as applications grow in complexity, and distance increases.

Before continuing, I'd like to pause and take back something I just said. I just wrote that when the web of real-time signaling is disrupted, "the illusion is shattered," but this is dangerously misleading. For it is the burden of this text to argue that in a very significant sense, the feeling of telepresence is *not* an illusion at all, or to be more precise that either the basic feeling of presence is *always some kind of illusion*, even in the normal everyday case, or if you don't want to count that feeling as illusory, the case of feeling the cup with my hand and feeling it with the telemanipulator are really, in the deepest sense, potentially on a par. I am arguing here for a kind of parity. Our sense of bodily presence is always constructed on the basis of the brain's ongoing registration of correlations. If the correlations are reliable, persistent, and supported by a robust, reliable causal chain, then the body image that is constructed on that basis is well grounded. It is well grounded regardless of whether the intervening circuitry is wholly biological or includes nonbiological components.

The less constant and reliable the correlations, however, the less willing the brain becomes to construct a body image to match. Even the basic biological body image is, surprisingly, hostage to such disruptive fortune. Certain neural malfunctions can play havoc with our sense of our own biological bodies, as in the condition of neglect in which a patient ignores, or disavows ownership of portions of her own body.[33] But as far as the technologies of telepresence are concerned, the major threats all depend on the timing and the nature of the signal exchanges that run *between* the biological brain, the human body, and the distant system.

One obvious threat is the disruption or delay of the signals themselves. You have probably had the experience of typing a document while logged onto a distant site or using a busy network. For a while, you feel in perfect command as the letters form on the screen just as you expect, but then it all slows down. You type but nothing happens, then suddenly it all comes in a flurry. This is even more unsettling if you are trying to backtrack to a certain spot or to amend a word or phrase. The disruption to the smooth flow of command-and-effect makes the system feel like an opponent, rather than a transparent medium through which to carry out a task. Time delays between operator action and systemic response (or associated sensory feedback) are the major source of "alienation" between teleoperator and robot/application. Such delays are most intrusive of all, of course, when the intention is to create a kind of technologically mediated physical interaction. When we reach out to touch another's hand, we expect no delay in feeling the hand for which we reach. When we swing the golf club, we expect to feel the resistance of the air and the impact of the ball pretty much at once. Inappropriate or unpredictable delays can rapidly torpedo any sense of ongoing physical interaction.

The brain itself confronts a related problem. When I reach for a nearby physical object, such as the coffee cup in front of me, my hand and arm move smoothly, thanks to the use of various kinds of sensory feedback from my bodily peripheries. Especially important here is the sense (known as proprioception) of how our own body parts are currently oriented in space. But it takes time for signals to return from the bodily peripheries to the brain—too much time, it seems, for the signals to be used to generate the smooth motions to which we are accustomed! To solve this problem, the brain uses a very neat trick indeed. It uses a special piece of neural circuitry known as a *motor emulator*. This is a little circuit that takes a copy of the motor signal to the hand (say) and feeds it into a neural system, which has learned about the typical responses from those bodily peripheries that are likely to ensue. The emulator is thus like a little local scale model of the real circuit. It rapidly outputs a prediction of the signals that should soon be arriving from the bodily peripheries, and these are then used instead of the real thing. This emulator-based feedback is then used for ongoing error-connection and smoothing.[34] Should the neural circuitry supporting this emulator function be damaged, smooth reaching is impaired, and jerkiness and oscillations result.[35] In this way the brain constructs its own little "virtual reality" in order to compensate for the temporal delays, which might otherwise impede smooth motor activity. The same trick is widely used in industrial control systems, for example, in chemical plants and reactors where you simply cannot afford to wait for correction signals from the target system before making delicate adjustments to the inflow of chemicals or materials.

This same strategy can be used to overcome some of the time-delay issues in teleoperator systems. Time delays of about 250 milliseconds and upward are enough to upset the feeling of delicate, continuous control that is associated with feelings of genuine presence. By inserting emulator circuitry into the system at the operator end, this effect can be offset. Kim and Bejczy developed, in 1993, a control system for "telerobotic servicing in space," which used what they dubbed a "predictive/preview display technique" so that the operator could see, at once, the predicted effects of her current commands. This was shown superimposed over the actual video footage returning from space.[36] The operator uses, in real time, what is in effect a prediction of the effects of her action on the distant target. The standard skeptical reaction—that this is replacing "real presence" with mere storytelling—needs to be rethought once we realize that our own brains, in guiding our daily actions, routinely create and exploit similar mock-ups!

The second potential source of user/device alienation is, of course, the telerobotic paradigm itself. Recall that in telerobotics what is communicated is a high-level command ("fetch the big rock on your left"), rather than a detailed motor sequence. When we *simply* issue high-level commands, we do not usually feel as if the devices (robots, sensors, waiters in a restaurant) are limb-like extensions of ourselves. Why is this? We saw earlier that our conscious mind often seems to issue high-level commands, which various neural "servo-systems" then translate into a full plan of motor action. Yet we do not usually suffer from the sensation that our bodies, while performing the action sequences determined by the zombies inside, are somehow less than full and proper parts of ourselves. This apparent asymmetry is in need of explanation.

Here is one possible story. In the case where my own detailed bodily motions are programmed by automatic neural subsystems, I retain a sense of *unfolding* and of *potential intervention*. By a sense of unfolding I mean a rich, if seldom foregrounded, array of sensory feedback (not just visual, but proprioceptive too), which keeps me apprised of the ongoing *details* of the action. By a sense of potential intervention I mean the knowledge that should things start to go visibly awry, I (my conscious self) can zoom in and place my hand and arm movements under closer conscious control.

These twin factors (ongoing feedback and the potential for further, more fine-grained intervention) instill in me the feeling that it is *me* acting, despite the fact that a lot of the problem-solving burden is borne by semiautomatic zombie subsystems.

Advanced telerobotics may need to provide similar resources if it is to combine "supervisory command style" interfacing with a richer sense of extended presence. One possibility involves what is currently known as "traded control." In a traded control system (e.g., Hayati et al., 1990) low level control of the robot can be delegated to an automatic system *or* taken over by the operator. Other variants include "shared control," in which certain operations are controlled by high-level commands while others are run in true teleoperator fashion. All these control modes (supervisory, traded, shared) can be mixed and matched while performing a complex task, perhaps bolstered by the use of emulation-based predictive resources.[37]

Reflecting on these varied resources, and on the extent to which the biological brain's normal interactions with body and world are mediated by a wide variety of similar tricks and ploys, it is hard to resist the general conclusion that these new technologies really *could* extend our sense of physical presence in important ways. Blake Hannaford, a professor of electrical engineering at the University of Washington in Seattle, sums it up like this:

> As robots and advanced user interfaces are connected to the Internet, we raise the possibility of the Internet connecting distant points in space with virtual, visual, aural and physical links. If the resolution of sensors and activators is high enough, and the bandwidth and latency adequate, we create "knots" or "ports" in space through which we can see, hear, touch and manipulate distant objects or people as though they were present. . . . What this will mean for the human . . . sense of presence is just beginning to be studied.[38]

Reshaping Presence

The transformative potential of the technologies of telepresence is enormous, but the precise shape of these imagined knots in space is still hard to determine. We should not simply assume that the most effective use of these technologies lies in the attempt to re-create, in detail, the *same kinds* of personal contact and exchange with which we are currently familiar. In fact, if we expect these technologies to deliver, at a distance, the very same kinds of sensory input and interactive potential that we encounter in "normal" daily life, they will almost certainly continue to disappoint. What if we instead allowed them to define *brand new niches* for genuine action and intervention?

The idea would be to allow the technologies to provide for the kinds of interactions and interventions for which they are best suited, rather than to force them to (badly) replicate our original forms of action and experience. After all, our single most fantastically successful piece of transparent cognitive technology—written language—is not simply a poor cousin of face-to-face vocal exchange. Instead, it provides a new medium for both the exchange of ideas and (more importantly) for the active construction of thoughts. We celebrate it for its special virtues, not as an impersonal, low-bandwidth, less rapidly responsive stand-in for face-to-face exchange.

This point is nicely made in a short piece by two Bellcore researchers, Jim Hollan and Scott Stormetta. The piece is called "Beyond Being There" and kicks off with an analogy.[39] A human with a broken leg may use a crutch, but as soon as she is well, the crutch is abandoned. Shoes, however (running shoes especially), enhance performance even while we are well. Too much telecommunications research, they argue, is geared to building crutches rather than shoes. Both are tools. We may become as accustomed to the crutches as the shoes, but crutches are designed to remedy a perceived defect and shoes to provide new functionality. Maybe new technologies should aspire to the latter. As they put it:

> [much] telecommunications research seems to work under the implicit assumption that there is a natural and perfect state—*being there*—and that our state is in some sense broken when we are not physically proximate. . . . In our view, there are a number of problems with this approach. Nor only does it orient us towards the construction of crutch-like telecommunications tools but it also implicitly commits us to a general research direction of attempting to imitate one medium of communication with another.[40]

Consider e-mail. E-mail is often used even when the recipient is sitting in the office next door. I do this all the time. My neighbor is a university colleague and for certain delicate, slow conversations, we much prefer a slow, asynchronous e-mail exchange. But e-mail is *nothing like* face-to-face interaction, and therein lies its virtues. It provides *complementary functionality*, allowing people informally and rapidly to interact, while preserving an inspectable and revisitable trace. It does this without requiring us both to be free at the same time. Cell phone text messaging has related virtues. The tools that really take off, Hollan and Stormetta thus argue, are those that "people prefer to use [for certain purposes] even when they have the option of interacting in physical proximity . . . tools that go *beyond being there*."[41]

Research into virtual reality (VR) has been, at times, a casualty of the crutches-not-shoes mind-set. So too has research into that branch of VR concerned with electronically mediated sexual contact: the subarea rather picturesquely known as "teledildonics."[42] In teledildonics research, data sensing "condoms" and dildo-like vaginal inserts communicate signals and motions between the genitalia of distant agents. In combination with bodysuits and head-mounted displays, the idea is to try to simulate many of the details of real bodily touch and sexual intercourse, and thus to re-create standard sexuality at a distance. Here too, the forward-looking theorist might also consider (more imaginatively and perhaps more successfully) using electronic means to expand, rather than simply inadequately reproduce, the normal range and repertoire of human touch and exchange.

Returning to mainstream (or vanilla as it were) VR, we may note that large amounts of such work are hostage to three distinct problems. The first is that perception is not passive. As we saw earlier, it will not be enough to present the eyes with a fully realized, rich three-dimensional (3D) scene if we cannot also in some way move and act within the scene itself. The second is that even if you add moving and acting, the day of full, multisensory, high-bandwidth, real-time, two-way interaction via telepresence remains distant. The third, and most important, is that even full telepresence, thus achieved, might be more of a crutch than a shoe. It might achieve precisely the goal imagined in Dennett's original thought experiment, of effectively "stretching out the nerves" so that we experience ourselves at a new location—but it would not expand the *types* of engagement we enter into, nor would it fundamentally alter our own experience of being in the world.

Yet the greatest potential of the technologies of telepresence, VR, and telerobotics may be transformative rather than replicative. It is not just a matter of (in effect) providing an electronic, information-based subway system so that we can move rapidly from place to place, avoiding the traffic and pollution! Rather, it is about expanding and reinventing our sense of body and action. Such reinvention, as Ramachandran's elegant experiments showed, comes surprisingly easily to brains like ours. The body image itself is highly negotiable, and the brain is plastic enough to learn to exploit whole new kinds of feedback loop and action-potential.

To get a sense of the kind of thing I have in mind, consider a familiar (and perfectly proper) cluster of objections to the project of total telepresence, meaning the attempt to simulate—using various new technologies—normal, daily, physical proximity by artificial means. This cluster of objections might be termed the "objections from depth and intimacy." Here is a sampling:[43]

- You are "at" a telepresence conference meeting when you observe a participant having some kind of seizure or heart attack. What can you do? You can dial an ambulance, to be sure, but if you cannot extend physical help, can you really count as "being there?"

- Conversely, if the others "present" cannot physically harm you, can you really count as "being there?"

- In deep physical intimacy, we are constantly touching the other, and responding to his/her touch. Given time-delays it seems unlikely that any tele-intimacy system could replicate this mutually modulatory touch.[44]

- And isn't there always a kind of "fixed depth" to our teledealings? In the daily world we can zoom in as much as we like. If we suddenly choose to order a pizza into the conference room we can all share the taste and smell of that very pizza. In telepresence, the extent of our mutual sensory involvement is always fixed in advance, by the specific channels and bandwidths available.[45]

- Most generally of all, what about that somewhat ineffable "sense of being in the presence of other people"? The philosopher Hubert Dreyfus (borrowing a term from Merleau-Ponty) calls this the sense of "intercorporeality" and suggests that "even the most sophisticated forms of telepresence may well seem remote and abstract if they are not in some way connected with our sense of the warm, embodied nearness of a flesh-and-blood human being."[46]

This whole cluster of objections is, however, compelling only insofar as we accept that the *goal* of telepresence is to replicate standard biological presence. The future, however, may turn out to be rather more interesting than that.

A really simple example is the LumiTouch, a prototype picture frame.[47] When one user touches the frame, the frame of a connected-but-distant picture lights up. If the distant partner sees this, and picks up her frame and squeezes, a feedback display area lights up on the originating frame, its color and intensity varying according to the force, location, and duration of the distal squeeze. Over time, two distant participants can learn to exchange a kind of private emotional language of touch using the device.

Back in 1993, participants in California and New York experimented by placing their hands inside a "datamitt" (informally known as the Data Dentata; see fig. 4.4) containing a very coarse array of touch sensors and actuators. Using the mitt, a hand squeeze could be executed in New York and felt in California, and vice versa. In this simple experiment, people reported a strong sense of personal contact despite the very low bandwidth of the connection. A more recent example, from the MIT Tangible Media Group, is "inTouch."[48] This system consists of two distantly coupled triple rollers mounted on a base. Sensors monitor the forces applied to each

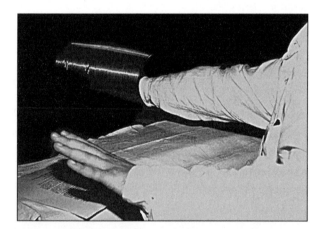

Fig. 4.4   The Data Dentata Installation. The user places her hand into an electrome-chanical device containing a binary switch, which allows transmission and reception of one bit of information sent via digital modem to a symmetric arrangement at the other end of an ordinary telephone line. A hand squeeze launched in California can thus be felt in New York, and vice versa. Such simple devices can create a strong sense of co-presence. Photo courtesy of Professor Ken Goldberg.

roller, transmit the data, and the same forces are locally re-created on the distant roller. Users feel as if they are touching a single object, each one applying her own forces and motions, and simultaneously feeling the forces and motions imparted by the other.

For something even more exploratory, we can visit the work of John Canny and Eric Paulos. Canny is professor of computer science at UC Berkeley and directs the 3DDI (3D Direct Interaction) project there, Paulos is a graduate student and co-worker. Together, they are working on a new form of computer-mediated interaction, which deploys real physical robots acting as personal representatives. The robots are called ProPs (Personal Roving Presence devices). Your ProP would be unique to you (rather like an avatar in virtual reality), but it would not look like you, so much as like a mobile cubist statue (see fig. 4.5). Each ProP would provide gaze-control, body-control, posture, and dialogue. Canny and Paulos's aim is to create ProPs that become transparent interfaces between a remote operator and their local contact: "Think of a ProP as the ultimate prosthetic: a full-body replacement . . . still fully under the control of a human being."[49]

In their writing, Canny and Paulos touch directly on two of the themes developed earlier. They stress the need for "transparent control" and interfaces that become invisible in use, and they note our amazing facility for "learning and living inside

Fig. 4.5 A ProP (Personal Roving Presence device), designed by John Canny and Eric Paulos. The device acts as your representative at a distant site and provides gaze control, body control, posture, and dialogue. Photo courtesy of Eric Paulos.

a different body."[50] Their hope is that over time, the human and her ProP become so well coupled that we learn to use the rather restricted range of ProP motions and displays to convey rich and subtle messages, much as skilled text messagers use that low-bandwidth resource to convey subtle emotional messages. Part of the idea is thus that a few relatively simple kinds of tele-interaction might yield a more robust sense of presence than a failed attempt to re-create the full gamut of human "intercorporeality."

The goal of Canny and Paulos's research is, they say, "not a human and robot hybrid but a new kind of embodied person."[51] This diagnosis is echoed by N. Katherine Hayles, a professor of English at UCLA. It is never, Hayles argues, a matter of "leaving the body behind." Instead, the technologies of telepresence and VR are about "extending embodied awareness in highly specific, local, and material ways that would be impossible without electronic prostheses."[52]

The larger lesson, then, is that embodiment is *essential but negotiable*. Humans are never disembodied intelligences; work in telepresence, virtual reality, and telerobotics, far from bolstering any mistaken vision of detached, bodiless intelligence, simply underlines the crucial importance of touch, motion, and intervention. In all the cases we have examined, what *matters* are the complex feedback loops that connect action-commands, bodily motions, environmental effects, and multisensory perceptual inputs. It is the two-way flow of influence between brain, body, and world that matters, and on the basis of which we construct (and constantly re-reconstruct) our sense of self, potential, and presence. The biological skin-bag has no special significance here. It is the flow that counts.

The deep and abiding importance of work in telepresence and virtual reality thus goes beyond the mere technological promise of new "knots in space." It goes, too, beyond the pragmatic, personal, and perhaps sexual attractions of exploring multiple forms of embodiment and variations of personal interaction. What we really stand to gain, I think, is knowledge about who and what we are. We learn that we are *essentially* active, embodied agents, not disembodied intelligences that simply manipulate or animate our biological bodies. We also learn—and this is the crunch—that the forms of our embodiment, action, and engagement are not fixed. New technologies can alter, augment, and extend our sense of presence and of our own potential for action. Even when they fail, when they reveal themselves instead as loud, abrasive, opaque barriers between us and our worlds, we learn a little more about what really matters in the ongoing construction of our sense of place and of person-hood. In success and in failure, these tools help us to know ourselves.

CHAPTER 5

What Are We?

Stelarc's
Third Hand

Hosei University, Tokyo, 1982. The man on stage has three hands. Two of them are his standard biological kit; the third is an electronic prosthesis. It looks like a somewhat rigid electronic shadow of the fleshy right forearm and hand. Built to those same dimensions, it is attached to the right arm (see fig. 5.1) and features grasp-release, pinch-release, and a 290-degree wrist rotation mechanism. The third hand is controlled by the man, via EMG signals detected by electrodes placed on four strategic muscle sites on his legs and abdomen.[1] In effect, the third hand is thus controlled by sending commands

Fig. 5.1 Performance artist Stelarc's Third Hand in action in Tokyo, 1980. Photo by S. Hunter, provided courtesy of Stelarc.

115

to these muscle sites, which act (via the electrodes) as a kind of relay center passing on the messages to the prosthesis. Since these muscle sites are not normally used for hand control, the third hand can be moved quite independently of the other two.

The Tokyo performance was an early outing. The performer, the Australian cyber-artist Stelarc, does not wear the device all the time. Nonetheless, over many years of use, he reports that he is

> able to operate the third hand intuitively and immediately, without effort and not needing to consciously focus. It is possible not only to complete a full motion, but also to operate it with incremental precision.[2]

In use today, the third hand is fluently integrated. He can use it for writing (see fig. 5.2), and it participates in many forms of deliberate action, often in close cooperation with its biological cousins. Stelarc does not feel that he "operates" the third hand. Instead, he simply uses it as he does the other two. It is like an occasional but fully paid-up member of his real body.

Stelarc is probably the most thoughtful, careful, and farsighted practitioner of cyber performance art alive today. For the last twenty years he has

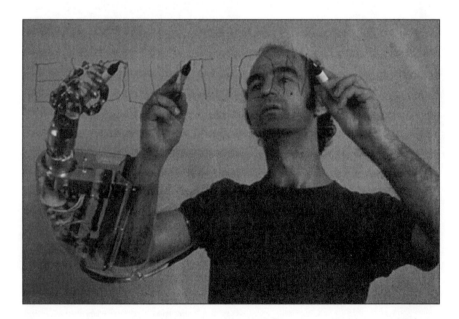

Fig. 5.2   Writing with three hands simultaneously, at the Maki Gallery, Tokyo, 1982. Photo by K. Oki, provided courtesy of Stelarc.

been delicately exploring the complex space of possible relations between body, machine, self, and agency. "As interface," Stelarc affirms, "the skin is obsolete. The significance of the cyber may well reside in the act of the body shedding its skin."[3] Yet this process of shedding, Stelarc simultaneously insists, does not herald the return of an outmoded notion of the person as a disembodied thinking thing. Rather, it invites us to explore a new realm of complex and multiple embodiment, with an associated expansion and enrichment of the subjective sense of self (fig 5.3). Stelarc's vision, like his performances, is complex and multilayered.

Counterpoint to the Third Hand, in the Stelarc repertoire, is what he terms "the Involuntary Body." Here other, sometimes distant, agents control Stelarc's biological body. The technology involves a six-channel touch-screen-interfaced muscle stimulation system. The system is operated using a touch-screen body display that delivers 0–60 volts to certain muscle sites (deltoids, biceps, flexors, thigh, and calf). At the higher voltage levels, this stimulation causes involuntary movements, which are nonetheless fairly smooth, as the voltage is delivered incrementally over several seconds.

Fig. 5.3 An extreme form of alternate embodiment! Exoskeleton is a jerky, stiff-jointed 600 kg machine that uses eighteen pneumatic actuators to drive its three degrees of freedom legs. The upper torso of the biological body controls the mode and direction of motion using magnetic sensors on the joints. In action in Bern at Dampfzentrale as part of the Cyborg Frictions performance in 1999. Photo by D. Landwehr, courtesy of Stelarc.

Imagine, now, a combined performance in which Stelarc's biological left arm is electronically stimulated by other people, while the right arm and the electronic prosthesis remain under Stelarc's own control.[4] On stage we see a man, sporting an extra, mechanical hand. Off to one side is a computer, with touch-screen interface, and a seated operator. You see the body on stage moving. The computer, you immediately suspect, must be controlling the mechanical hand. But you are quite wrong. The mechanical hand is under the voluntary control of the man on stage, courtesy of the electrodes on his legs and abdomen; one biological arm is under the control of the computer and its seated operator, via the voltage-bursts delivered by the muscle stimulators. The locus of voluntary control that, to all intents and purposes, is the person—Stelarc—has been expanded to include some nonbiological parts and circuits, and it has been contracted, with parts of the biological body now dancing to the tune of another's desires.

What Stelarc is doing, with wit, intelligence, and a keen dramatic sensibility, is extending his own nervous system into nonbiological space, while allowing other people's nervous systems to invade, manipulate, and parasitize aspects of his biological body. In fact, even this may paint too simple a picture, for what Stelarc ultimately cares about is neither simple extension nor simple contraction. Instead, he cares about the possibilities for new kinds of collaboration, skilled action, and intimacy that cyborg technologies provide.

Stelarc's vision is positive and liberating. The kinds of self-exploration the technologies permit will, he hopes, enhance and expand our sense of our own presence and our awareness of, and intimacy with, others. Is this, however, just effective, thought-provoking theater, or will new technologies truly alter and expand our sense of presence, body, agency, and control? If the latter, will they really do so for the better, or is there something nasty lurking under those biomechanical rocks?

It is too soon to say. As with all new technologies, the social and personal impact of bioelectronic interpenetration is difficult to predict. Even were the shape of the actual technologies clear, the ways in which they will become most widely used and incorporated into our daily lives is elusive. This is a general truth about instrumentally mediated societal change and in no way unique to the cases at hand. Donald Norman makes the point colorfully, noting that the inventors of the phonograph originally imagined that its main use might be for public demonstrations with paid admissions,[5] that the telephone was to be used to transmit daily news to the populace, who would gather around a single outlet,[6] and that the idea of anyone finding a use for a computer in the home was, at one time, considered laughable.[7] Perhaps all that can be said, with real certainty, is that the ideas and possibilities that Stelarc is exploring are not just theater.

Mind Control

Take the Third Hand. The Third Hand is only indirectly controlled by neural signals from Stelarc's brain (commands to move the hand are routed via commands to move specific muscles on his legs and abdomen). Even more direct forms of "mind control" are already the topic of much ongoing scientific research. We already encountered, in chapter 4, the work of Miguel Nicolelis of Duke University in Northern Carolina.[8] Nicolelis and his team used a kind of "neural wire-tap" involving ninety-six wires implanted into the frontal cortex to gather data about the correlations between patterns of neural signals and specific motions. Signals from the brain of the North Carolina owl monkey were then used to control the operation of a robot arm six hundred miles away in the MIT Touch Lab.[9] The neural commands were successfully translated into actual motions of the remote robot arm, which mimicked the full range of motions of its biological template.

The team's long-term goal is to develop a practical interface, which would allow a paralyzed human to control an artificial limb simply by willing the limb to move in the usual way. An onboard computer wired into the person's brain would detect the neural signatures of different motions and use them to control the robot limb. Moreover, as the work with the owl monkey effectively demonstrates, the same technique could be used to control the motion of more distant apparatus. Such techniques, combined with real-time visual and force-feedback loops (see chapter 4), begin to open up the possibility of real remote presence—piano recitals and delicate operations performed "live" at a distance.

From the patient's point of view, the robotic limb would respond to his will as directly as its real, biological counterpart. Stelarc, using a much simpler system, already reports having the feeling of immediate, effortless control over the Third Hand. And this despite the fact that the neural commands, in Stelarc's case, are nonstandard: Stelarc must activate muscle groups in his legs and abdomen to move the hand. What we are witnessing, in Stelarc's fluent performances, is yet more evidence of the remarkable capacity of the human brain to learn new modes of controlling action and to rapidly reach a point where such control is so easy and fluent that all we experience is a fluid, apparently unmediated mesh between will and motion. This fluid mesh is familiar to us all and not only from our experiences of bodily control. The expert car driver, golfer, tennis player, or video games player, will likewise have reached a point where aspects of the apparatus (the clutch pedal, the racket) become transparent in use.

Direct neuroelectronic interfaces are, of course, immensely appealing as technologies of profound human machine integration. A much-publicized recent success involved the use of the actual brainstem of a fish (a lamprey, which is an aquatic vertebrate somewhat like an eel) to control the motions of a standard, commercially available robot. The robot, known as a Khepera, was a small, round, wheeled device roughly the size and shape of a double stack of hockey pucks. The word Khepera means, as I recall, dung beetle. These Swiss-built robots are now a standard platform for simple robotics research; I had two in my robotics lab in St. Louis. Such robots are usually controlled either by a programmable onboard chip or by signals from a nearby PC. What Ferdinando Mussa-Ivaldi (Northwestern University) and Vittorio Sanguineti (University of Genoa) together with colleagues at the University of Illinois did was to replace the PC with a small piece of genuine lamprey neuroanatomy.[10] They removed the brainstem and a section of spinal cord from a lamprey, keeping the tissue alive in an oxygenated salt solution. They then isolated, within this preparation, a group of large nerve cells (knows as Myler cells) whose biological role is to help the lamprey orient in response to sensory stimuli. The Khepera robot's onboard light sensors were wired to the lamprey "brain," which was then able to control orienting responses in the robot body. Behaviors thus produced included following the light, avoiding the light, and turning in a circle.

Another animal-machine hybrid is the truly strange, half-virtual, half-biological creature built by Steve Potter.[11] Potter, a scientist at California Institute of Technology, works on silicon-neural interfacing and has developed a stream of potent techniques aimed (in part) at the production of hybrid neural and artificial computing systems. A flagship project, now funded by the National Science Foundation, aims to build a neurally controlled artificial animal. This is a simulated creature, living in a virtual world but controlled by real (rat hippocampal) neural tissue cultured in vitro on a glass-and-electrode sheet: a setup that allows Potter and his colleagues to watch and record as the neural elements guide the virtual creature. Such studies may provide the bedrock understanding needed to interface human neural elements with silicon circuitry to control artificial prostheses and other devices.

The mention of "other devices" is significant, for these technologies promise much wider applications too. Niels Birbaumer and a team of researchers at the University of Tübingen, Germany, have enabled a paralyzed patient to move a cursor on a computer screen using what Birbaumer calls a TTD (Thought Translation Device).[12] This is a noninvasive setup in which electrodes on the patient's scalp detect changes in slow cortical potential (SCP)—the electrical signature that typically precedes action. Different amplitudes of SCP were linked to different types of cursor movements. After allowing a patient to learn by trial and error how to generate these SCPs at will, the patient was able to pick out letters to form simple messages, just by using this controlled neural activity. As it stands, this way of moving a cursor to pick among options is a slow and painstaking process, but the idea is to add software innovations, which then allow the simple ("one mental click") selection of complex sets of phrases and commands. In this way, the "thought control" of simple everyday appliances (lights, TV, garage doors, etc.) could be facilitated. The potential applications of various kinds of thought control technology thus far outrun the arena—important though it is—of rehabilitation and recovery.

An alternative strategy is to build the thought control apparatus directly into the brain itself. A team of researchers led by Roy Bakay, a professor of neurological surgery at Emory University in Atlanta, Georgia, have successfully enabled a paralyzed stroke victim to control a cursor using two neural implants. The implants consist of two tiny glass cones surgically introduced into the patient's motor cortex. These cones are coated with special neurotrophic chemicals extracted from the patient's knee, which prompt nerve growth. These chemicals help the cortical neurons grow into the glass cones and then attach themselves to small electrodes inside. The implants transmit signals to an amplifier (worn in a cap), which relays them to the computer. The patient then, with effort and practice, is able to learn to use neural signals to control simple cursor movements. This is a bit like sending a text message using your cell phone but forming the letters in the message by mental effort alone. After a while, this activity becomes, to quote Bakay, "second nature."[13]

At first, though, in order to get the cursor to move, patients need to experiment with their own motor signals. The implants, remember, are lodged in the motor cortex—the area controlling bodily movement. The nerves that grow into the electrodes are thus likely to carry signals, which normally would participate in the control of such movements as the raising of an arm, a leg, or the wiggling of a finger. To successfully move the cursor by thought, the paralyzed patient first tries to will the motion of various bodily parts. When such efforts yield a signal, which the computer hears, a buzzer sounds so the patient knows to concentrate on that particular kind of thought.

This may sound initially disappointing. It isn't, after all, the kind of thought control we once read about in science fiction. In science fiction, it is (often) the *contents* of the thoughts that seem to cause the objects to respond. The heroine thinks "ashtray" and here it comes. She thinks "ray gun" and it flies right into her hand. But to move the cursor to the right, the stroke victim cannot just think "cursor, right." Instead, it is like willing a bodily part to move. The cortical implant, then, is really not so different from Stelarc's system for controlling the Third Hand, except that the implant allows the muscle control signal to be intercepted and exploited earlier in the causal chain—*before* the muscle itself responds (or fails to, in the case of the paralyzed patient).

Nonetheless it is worth pausing to ask ourselves whether the type of thought control thus achieved is in some way *essentially* different from the way a normal brain might control a bodily member. At first it seems to be quite different, since to get the cursor to move, the subject may need to *will* something different, for instance, "lift the leg." But notice that after a while, the mental reflex becomes second nature: when you want the cursor to move, you just will it to do so, exactly as you might will your own leg to move. Stelarc reports exactly this experience with his control over the Third Hand. The expert snooker player feels the same way. She does not consciously intend such and such movement of the cue. Rather, she intends to put the last red in the center pocket and to spin back for the black. To repeat a now-familiar tune, it is when aspects of body or external tools become transparent in use so that our intentions "flow through" the tools to alter the world, that we feel as if we directly control the limbs, or tools, in question, that we begin to feel as if they are a part of us. After sufficient practice, the kind of thought control exerted by the paralyzed patient over the cursor, by the normal subject over her biological arm, and by the snooker ace over her trusty cue all look to be pretty much on a par.

Well, almost. To move the snooker cue you must first move your arm, even though you need not consciously intend to do so. But to move the arm itself or the cursor, after practice with the cortical implant, you need not first move anything else. So the sense in which the cursor is directly controlled by your neural activity is about as strong as can be.

Someone might still insist there is a significant difference. Surely, they might say, the neural activity that normally causes you to raise a leg actually *means* "raise that leg," whereas the neural activity that drives the cursor really *means* (let's say) "wiggle your left finger," even though, courtesy of the new circuitry, it now causes the cursor to move. To get to the bottom of this rather profound mistake would require a long philosophical detour, but I invite the reader to try the following thought experiment. Imagine you are an infant, above whose crib dangles an attractive mobile. You want to touch it, but you do not yet know how to issue the correct motor commands to do so. Your brain, however, generates many bursts of essentially random neural activity. Some of these bursts seem to move your hand closer to the target. After a while, you learn how to generate this kind of neural activity at will, and hence how to control your own limbs so as to carry out your project.[14]

If this is a good picture of how we first learn to control our own bodies, and it seems it is (see note 14), then there is a sense in which the neural activity that yields some desired result, like a certain kind of arm motion, counts as meaning "move that arm" only because it is the kind of activity that brings about just *that* kind of result. It is not that the neural signal, in some independent sense, means "move that arm." Rather, it means what it means because of what we can reliably use it to bring about. If, later in life, we learn to use some pattern of neural activity to control the cursor, we are simply doing more of the same. The pattern of activity that promotes a specific cursor movement should count as having the content "move the cursor to the right" (or whatever) in as rich and real a sense as it once counted as meaning "wiggle the left finger." In fact, it could come to mean both things simultaneously, were both chains of effect functional and intentionally exploited.

In short, and despite initial appearances, I claim that the kind of thought control, which cortical implants make available, is thought control *in precisely the same sense* as is implied when we say that our thoughts can make our fingers move, and so on. It is different, indeed, to that science fiction scenario in which I rehearse a kind of mental sentence, saying "come, ashtray" or "to me, raygun." But that is not, in any case, our usual mode of neural control. When I want to go for a run, I don't say to myself "go feet."

Work on neuroelectronic interfaces is also being pursued in the opposite direction. The work just described was concerned with translating neural impulses into action, but an equally important complex of projects targets the transformation of sensory inputs into neural patterns. We already encountered, back in chapter 1, some quite sophisticated work involving cochlear implants for the deaf. Certain artificial vision systems, likewise, aim to bypass damaged optical resources and transmit visual information directly to the blind person's visual cortex. One successful prototype system (fig. 5.4) is called the "Dobelle Eye."[15] A small TV camera and an ultrasonic distance sensor are mounted on a pair of sunglasses, connected by cable to a portable "fanny pack" computer worn on a belt. The computer integrates distance and visual information, and transmits a signal to a sixty-eight-electrode array implanted on the surface of the visual cortex. After a period of training and experimentation, one patient in a recent study (a 62-year-old male, blinded in an accident at the age of 36) was able to

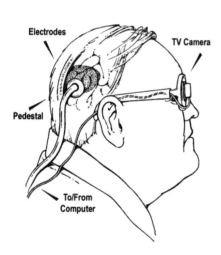

Fig. 5.4   A prototype artificial vision system, the "Dobelle Eye," bypasses damaged visual resources sending camera-and-computer generated signals direct to a sixty-eight-electrode array implanted on the surface of the blind person's visual cortex. Illustration courtesy of the Dobelle Group.

read two-inch tall letters at a distance of five feet and began to negotiate new and unfamiliar environments. In one especially suggestive follow-up experiment, the wearable TV camera was replaced by a variety of different packages, including one that linked his cortex directly to commercial TV (a terrifying thought this!), one to the internet, and one to a text-manipulating computer program. At moments such as these, the cyberpunk future seems unexpectedly close at hand.[16]

The basic idea is by no means new. In 1972, Paul Bach-y-Rita pioneered the use of TVSS (Tactile Visual Sensory Substitution).[17] This was a device worn on the back but connected to a camera worn on the head. The backpack consisted of an array of blunt-ended "nails," each nail activated by a region of pixels in the coarse visual grid generated by the camera. A more recent descendent of this device (fig 5.5) uses a much smaller, electrical stimulatory grid, worn on the person's tongue.[18] Fitted with such devices,

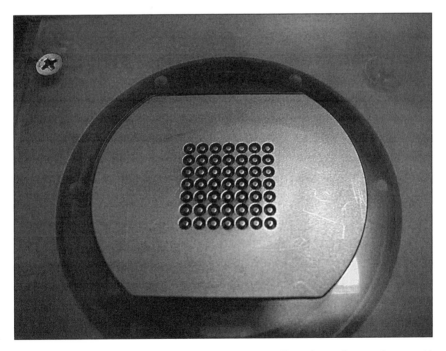

Fig. 5.5   Another artificial vision system. Here, a small 49-electrode stimulatory grid (7 × 7 electrodes spaced 2.54 mm apart) is worn on the tongue and receives input from a head-mounted camera. After practice, patients begin to sense objects in front of the camera and report visual-like sensations. Photo courtesy of Professor Paul Bach-y-Rita.

subjects report that at first they simply feel the stimulation of the bodily site (the back or the tongue). After extensive practice, in which they actively manipulate the camera while interacting with the world, they begin to experience coarse quasi-visual sensations. After a time, they cease to notice the bodily stimulations and instead directly experience objects arrayed in space in front of the camera. If the camera input, for example, presents a rapidly approaching object (signaled by a rapid expansion of the image in the TV camera, which translates into rapidly expanding activity of the tactile grid), the subject will instinctively duck, and in a way appropriate to the perceived threat.

The lesson, once again, is that our brains are amazingly adept at learning to exploit new types and channels of input. To be sure, more direct cortical interfaces will have many advantages, notably the potentially high bandwidth of information transmission. What really matters is simply the provision of a reliable and detectable array of incoming signals properly correlated with both the subject's own self-directed exploratory activity and with the changing states of the world. The human eye provides one such complex of information, the TVSS grid another, and the direct cortical interface yet another.

Defense agencies, as always, are keenly attuned to the potential of such research. DERA (Defence Evaluation and Research Agency), the British government's technological think tank, is working on the "Cognitive Cockpit"—an attempt to allow fighter pilots to operate certain control systems by direct neural activity or by gaze-direction (just looking at the relevant control).[19] Such pilots could also wear clothing that monitors their physical state and could cede control to an autopilot in case of a medical emergency. The pilot's brain and body are to be woven into the fabric of the cockpit control system. If this sounds off-putting to you, imagine being woven into the control system of a new Jaguar convertible (well, it works for me).

Putting all this together with the work reviewed in previous chapters, a pattern emerges. We are moving toward a world of wired people and wireless radio-linked gadgets. Infrastructure technologies (such as the wireless voice and data transmission system Bluetooth) will systematically liberate our domestic and office devices from cable-based communications links. The devices themselves will be dedicated yet semi-intelligent, able to receive and broadcast information among themselves; and some of the human brains and bodies embedded in this matrix will probably use implant technologies to communicate directly with some of the gadgets.

It is worth recalling, in this larger context, the speculative self-experimentation of Kevin Warwick, mentioned in chapter 1. Warwick, a UK-based professor of cybernetics, used an implant connected to nerve bundles in the arm to pass signals to and from a computer. Warwick planned to replay the signals back into his own nervous system and that of his wife (via a similar implant). The true value of such experiments in biocomputational communication may lie in the many ways we could slowly learn to use such new channels to control and coordinate whole new kinds of activity and machinery. Warwick's experiments can be extended (and Warwick is indeed working on this) in order to open a usable channel between the nervous systems of two individuals. Once such a new channel is open, there is the possibility that two brains might together learn how to use this novel resource to coordinate some joint activity such as dancing. Such scenarios, as we saw before, should probably not be depicted in terms of telepathic mind reading. Instead, the question is whether the brain's considerable plasticity might allow it—when presented with such a new resource—to learn how to directly influence the motion of another body. As an aside, it is interesting to consider the question of whether there might ever be *so much* influence, running in both directions, that it becomes proper to think of one mind with two bodies rather than two minds, each of which exerts some modest control over the other's body. Our own two brain hemispheres clearly enjoy enough dense communication to count as a single mind controlling a single body. Impair those channels (principally, the corpus callosum), and it often looks more like the opposite of the case we just imagined: two minds fighting for the control of a single body.[20]

Donald Norman notes that we humans, especially the younger ones, are amazingly good at acquiring new kinds of essentially arbitrary control and communication skills.[21] We can learn to speak by—incredibly!—puffing air through our vocal cords. We can learn to write, to touch-type, to fly planes and drive cars, and to play clarinet, oboe, and violin. Put this ancient capacity together with new direct electronic channels into the brain, perhaps even directly linking my brain and yours, simmer for long periods of intense practice, and who knows what new skilled forms of interpersonal and neuroelectronic harmony may emerge? As Norman put it:

Suppose we tapped a fast, high-bandwidth nerve channel . . . maybe we could tap into the spinal cord or into the nerves that go to and from a hand. Suppose we hooked up a high bandwidth channel that sent and received neural impulses through this tap. At first, they would simply lead to peculiar sensations and jerky, uncontrolled movements of the body. . . . But I suspect that if we undertook daily training exercises—a few hours a day for, oh, ten years— who is to say that we couldn't train ourselves to communicate?[22]

Norman's vision combines elements of the neural-cursor-control strategy with elements of Warwick's speculative experimentation. Instead of trying, ponderously and improbably, to rig our machines to "read our minds," we simply open a new channel and allow our brains to learn to control the machines. Were we to open a direct channel between my brain and yours, who knows what new modes of communication, control, and intimacy we might achieve? Commenting on such a vision, Norman notes that the upshot is not the re-creation or facilitation of existing forms of communication, control, and experience; rather it is the opening up of brand new horizons: new worlds of human-machine and human-human communication and interaction. What might such a world be like? Let's peer into the diary of a possible inhabitant of a future, but very possible, world.[23]

Day One.
I live and work in a world animated by invisible spirits. Or at least, that's certainly how it seems. My house, my automobile, and my office are all constantly aware of my needs and movements. My refrigerator knows when I am running out of milk (and it orders more). My car knows what the weather is like and begins to de-ice itself as soon as I fill the dedicated car-beverage container with hot coffee. My office knows when (and where) I have parked my car and alerts my clients (and coffeemaker) accordingly. Even the clothes I am wearing are part of this web of intercommunicating support systems. My shirt monitors my heart rate, temperature, and mood; it talks to the room and car when things look dicey. None of this, of course, is compulsory. I can turn it all off if I want to. But when it is up and running, the world seems a much less troublesome place. (Sometimes, I disable the technologies to experience that state of nature that was so common in the twentieth century: the state where technology was untamed, and

each individual had to fend for herself, pressing buttons and begging the machines to work.)

Day Two.

I really needed to spend time with Karen today. So in the morning, after a long, sultry chat, I telemanipulated her arm and hand (she is still away in California), and she telemanipulated mine. It was all a bit jerky, but the sensation of being touched by her is worth the effort. This technology still has a way to go. I recently read, however, of a pair of dancers who are really taking this stuff somewhere. While they dance, each one controls half of the other's body. It must take a lot of practice, but the feeling sounds strange and intimate.

Day Three.

Seems I have a problem. One of my software agents (my so-called Mambo Chicken Bot,[24] which has been learning about, and contributing to, my taste for the weird and exotic for three and a half decades, coming online when I was five and first fell in love with astrophysical oddities) is temporarily disabled. In fact, it has been out of action for a few months. I only found out today when a diagnostic web-surfing Bot e-mailed me the bad news. But I had been feeling unusually flat and uninspired for a while, so I should have guessed that something was wrong somewhere in my distributed cognitive web. The Mam-Bot's occasional unexpected inputs (I call them my inspirations) really do help. Well, at least it was only the Mam-Bot and not something more central. What must it be like to wake up one morning and find your most important Bots compromised by some horrible accident? It must be like becoming a child again.

Day Four.

I was animating the car (it's one of those new models that weaves the driver right into the road management system) when my neurophone signaled me. I suddenly remembered the old extrinsic cell phones of my mother and father—chunks of metal and plastic that you had to actually carry around! The neurophone is interfaced directly to my cochlear nerve, and the microphone in my jaw is sensitive enough to allow me to merely mime the words if I am in a public place. Lately, I forget that the ability to phone other people is actually a technological aspect of my being at all; it just seems like something I can do, like shouting someone's name.

Anyway, the call was to tell me that my new arm is ready to ship. It will take a little training to get the thing calibrated. But after a few weeks of practice, I'll be able to use it as effortlessly and as fluently as my biological arm. The new limb will be wired to electrodes that detect activity in my cerebral cortex. It's a cool prosthesis, since I've ordered some extras, such as a sixth finger (which, once I learn to control it, might help my guitar playing: who knows?), a lock-in putting grip (still illegal according to the PGA, but what the hell), and a built-in fingertip infrared camera capable of directly stimulating my optic nerve to allow me to "point and see" in the dark. In some ways, I feel lucky to have lost one arm all those years ago. Until legislation catches up with reality and allows unimpaired citizens to purchase and fit prosthetic enhancements, I really have an edge! (Of course, the law is totally confused here—the neurophone is surely an aural prosthesis of some kind, yet Congress admitted it without a second thought. And what about the car itself? Isn't it a kind of prosthesis now that it is open to such direct and effortless neural control?)

Soft Selves

Do you feel an identity crisis looming? Where, in this increasingly dense biotechnological matrix shall we locate our (human?) selves? The question can quickly confound, since the notions of self and identity are notoriously elusive. What is the self anyway? Does it make sense to even try to locate it? The philosopher Daniel Dennett offers the following formula. Control, says Dennett, is the ultimate criterion: "I am the sum total of the parts I control directly."[25]

In bringing control to the forefront, Dennett successfully captures one important feature of our experience of self. Suppose you are in an amusement park, in the hall of mirrors. You see a hand and arm reflected. Is it *yours*? The natural way to tell, Dennett notes, is to try to move it. You don't just ask whether it *looks* like your hand, you experiment by willing it to move in specific ways.

Of course, I can move *your* hand, too (I can grab your arm and shake it, for example), but this is a very different kind of experience. When I move my hand, I do so without needing to first locate it by vision, or to intentionally will anything else to move in order to make it move. For example, to move the table, I push it with my hands; but to move *my hands*, I don't need to push anything. This experience of direct responsiveness is a major factor in the creation of our sense of bodily presence. In his account of this sense of immediacy, Jonathan Glover, a British philosopher, suggests that

> if signals could be sent from my nervous system to receptors in physical objects detached from my body, so that I could move those objects in the same direct way I can move my arms, it might be less clear that I stop where my body ends. These doubts would be even stronger if sensory signals could be sent back, enabling me to "feel" things happening in the detached objects. We might then say that I extend beyond my body, or else we might treat these objects as free-floating parts of my body.[26]

Notice that any object *attached* to the biological body, like Stelarc's Third Hand, is automatically able to send such signals back, courtesy of the biological network of force sensors known as the haptic system. The capacity to feel the road with the cane is an example of this system in action.

The notion of "direct control" is thus meant to rule out the case where we must first control our own bodies and, using them as our instruments, affect something else. Stelarc's Third Hand, when attached and in use, is part of Stelarc himself in just this sense. The fact that Stelarc must control the hand by first contracting muscles in his legs and abdomen may seem to argue against this, but remember that after a while Stelarc does not experience the control structure that way. Instead, he simply wills the hand to move, and it moves. The fact that this involved a causal detour is unimportant. In a similar fashion, the "direct" cortical control studied by Roy Bakay actually requires the patients to issue neural signals that would normally move specific body parts but that now move the cursor backward and forward. The infant, likewise, must learn to control its limbs by finding which neural signals reliably issue in desired actions or outcomes. The relation between a movement and a neural signal is—from the agent's point of view—always somewhat arbitrary.

By linking the conception of the self to a conception of whatever matrix of factors we experience as being under our direct control, Dennett makes ample room for truly hybrid biotechnological selves. The most basic notion of the self, on this model, is simply the plastic, multiply negotiable sense we have of our own physical presence in the world. This sense is determined by our experiences of direct control—experiences that provide the kinds of statistical correlation between motor signals and sensory feedback, which Ramachandran (chapter 3) showed were sufficient to cause rapid changes in our sense of our own embodiment. The human self has however, another dimension. I think of myself not just as a physical presence but as a kind of *rational* or *intellectual* presence. I think of myself in terms of a certain set of ongoing goals, projects, and commitments: to write a new paper, to be a good husband, to better understand the nature of persons, and so on. These goals and projects are not static, nor are they arbitrarily changeable. I recognize *myself*, over my lifetime, in part by keeping track of this flow of projects and commitments. Others, likewise, will often recognize me as a unique individual, not (or not only) by recognizing my physical shape and form but by recognizing some distinctive nexus of projects and activities. Some writers speak here of the *narrative* self—the self identified by a story told both to ourselves and others, and told both by ourselves and others.[27]

This narrative self, I want to suggest, may be a biotechnological hybrid in a different, even a deeper, fashion. The narrative self is a self built out of our own and others' conceptions of our projects, capacities, possibilities, and potentials. Can we really suppose that it would make no difference, to the "I" thus identified, to find itself moving, thinking, and acting in a more highly biotechnologically integrated world? In a world where dedicated software agents constantly search the web for items of special interest and for new opportunities to carry forward the projects dearest to its heart? In a world where the capacity to use certain devices and software packages is as fluent and direct as the capacity to move my own biological body?

Fitted with a shiny new prosthetic arm that allows me to lift more weights than I could before, my sense of what I can do must rapidly alter and catch up. Fitted with a cochlear implant that cures my deafness and, as a kind of added extra, allows me to hear sounds in ranges that most adult humans cannot detect, my core sense of my own auditory potential again changes. Accustomed to the now automatic and unreflective use of, perhaps, a retinal display that allows me to rapidly and invisibly retrieve information from a linked database, it seems less and less clear where what "I" know ends and what "it" (the technology) makes available starts.

Imagine a twist on the earlier example of knowing lots of facts about American women's basketball. Instead of storing all those facts in your head, suppose that you now deploy a kind of heads-up display that provides instant access to the main performance statistics of key players over the last twenty years. The display might be delivered by eyeglasses or courtesy of a wireless implant sending signals directly into visual cortex, rather like the new-generation cochlear implants described in chapter 1. Either way, the system is set up so that the visual sighting of a player's name, or the auditory pickup of that name, or simply mouthing the name, activates a kind of augmented reality visual overlay displaying key facts and figures. Imagine, too, that the system is fairly flexible, allowing you also to start with categories (for example, "three-point field goal percentages in the year 2000") or with specifics ("players with three-point field goal percentages of .350 or above") and then retrieve information accordingly.

Over a period of use, you become so accustomed to this easy, on-demand access that the retinal display becomes transparent equipment. As a result, you may feel as if you simply *know* (always) which of any two players had the best three-point field goal percentage in any given season. Would you be wrong to feel that? Perhaps not. True, your access to these items of information depends on the proper operation of the technology, but your knowing other things depends, equally, upon the proper operation of parts of your brain. In each case, damage and malfunction is a possibility. And true, you need to actively retrieve the information before it becomes available to your conscious awareness. However, knowledge stored in long-term biological memory is in just the same boat, awaiting activation by some kind of retrieval process to poise it for the control of verbal report and willed action.

This is not to say that there are no interesting differences. For example, knowledge stored in long-term biological memory is open to all kinds of subterranean processes of integration and interference with both old and newly acquired knowledge; neutrally stored information is fluently accessible by an amazing variety of routes and in a wide variety of situations. Nonetheless, the simple feeling of "already knowing" the answer to a question as soon as it is asked is surely the knowledge-based equivalent of the more generic notion of "transparent equipment." Easy access to specific bodies of information, as and when such access is normally required, is all it takes for us to begin to factor such knowledge in as part of the bundle of skills and abilities that we take for granted in our day to day life. It is this bundle of "taken-for-granted" skills, knowledge, and abilities that structures and inform our sense of who we are and what we know.

Most but not all theorists will agree that there is a genuine (by no means sharp and "all-or-nothing") distinction between those things of which I am consciously aware and those things of which I am not.[28] Right now, for example, I am conscious of the page in front of me, of the glare of my desk lamp, and of the difficulty of formulating this particular thought! I am not, however, conscious of all the complex low-level visual processing (for example, the parallel processing of multiple differential equations) that supports and makes possible my conscious visual awareness of the page and the glare. Nor am I conscious of whatever complex internal machinations underlie my sudden sense that I am here tiptoeing into difficult and dangerous territory. Certainly, at any given moment, not *all* the cognitively important goings-on in my brain are present as *contents* of my current conscious awareness. That is why we sometimes find thoughts and ideas simply "popping up in our heads"; they are the intrusive conscious fruits of some ongoing, subterranean, nonconscious information processing.

It is impossible to underestimate the significance of these nonconscious cognitive processes in the determination of the mental character of a persisting and identifiable thinking being. We must reject the seductive but ultimately barely intelligible idea that we (as individual, thinking things) are nothing more than a sequence of conscious states. Such a view obscures the mechanisms responsible for the kind of *cohesion* and *continuity* that we naturally associate with the idea of a single self or mind persisting through time.

If you don't believe me, try the following experiment. For thirty minutes, keep track (as far as you can) of the contents of your conscious awareness. Unless you are totally engaged in a single all-absorbing task, you will probably end up with a sequence of often-unconnected thoughts. A feeling of hunger, a thought about consciousness and the self, a worry about a forthcoming lecture, a glimmer of sexual arousal, a pang of anxiety, an urge to write to an old friend, another thought about the self (a good one but where did it come from?), the strong desire for a cup of coffee, and so forth. This sequence of conscious contents is highly varied in type and radically discontinuous in content. Themes persist and whole trains of thought are, sometimes painfully, birthed, but the true principles of continuity lie largely underground.

Taking all this nonconscious cognitive activity seriously is, however, already to take the crucial step toward understanding *ourselves*, in the very deepest sense, as biotechnological hybrids. The relations between our conscious sense of self (our explicit plans and projects, and our sense of our own personality, capacities, bodily form, location, and limits) and the many nonconscious *neural* goings-on, which structure and inform this cognitive profile, are pretty much on a par with the relations between our conscious minds and various other kinds of transparent, personalized, robust, and readily accessed resources. When those resources are of a recognizably knowledge-and-information based kind, the upshot is an extended cognitive system: a biotechnologically hybrid mind.

Confronted with this bold proposal, many people feel deeply uncomfortable. How, they ask, could something to which I have so little access count as in any way a part of *me*? To see how this could be so, it helps to reflect that even in the case of our own biological brains, the conscious self is in direct control of much less than we think. Not just the autonomic functions (breathing, heartbeat, etc.) described in chapter 1, but all *kinds* of human activities turn out to be partly supported by quasi-independent nonconscious subsystems. This is no surprise, I am sure, to any sports player: it doesn't even *seem*, when playing a fast game of squash, as if your conscious perception of the ball is, moment-by-moment, guiding your hand and racket. Nor should it come as a surprise to artists and scientists, who are often painfully aware that the bulk of their own creative activity is subterranean and nonconscious.

None of this seems to bother us unduly. All that seems to matter, really, is that the conscious self has a broad sense of what the entire situated and embodied agent can and can't do. It is this sense that enables us to plan our lives and projects. The conscious mind, on this model, emerges as something like a new-style business manager whose role is not to micromanage so much as to set goals and to actively create and maintain the kinds of conditions in which various contributing elements can perform best.[29]

The skill of successful *self*-management is thus the skill of knowing how to exercise rather indirect (softly, softly) forms of intervention and control—what Kevin Kelly nicely dubs "co-control."[30] Instead of handing down detailed blueprints and game plans, the conscious mind/manager acts like a nanny or coach, trying to nudge and cajole her charges into giving their best.

A common objection, at about this point, goes something like this: even if external, nonbiological elements do sometimes help us, quite profoundly, in our problem-solving activities, still isn't it always our biological brains that have the final say? The mental buck stops there. The brain is where *I am* because the brain is the controller and chooser of my actions in a way this other stuff (software, pen, paper, Palm Pilot) is not. And *that*, it may be suggested, is why the nonbiological stuff should not count as part of the *real* cognitive system, and why our minds are not hybrids built of biological and technological parts. Human minds, so the obvious objection goes, are good old-fashioned biological minds, albeit ones that enjoy a nice wraparound of power-enhancing tools and culture.[31]

This sounds sensible and proper, but only until we turn up the magnification on the biological brain itself. Notice first that many processes involved in the selection and control of actions *are* routinely off-loaded, by the biological brain, onto the nonbiological environment. Think of the knot in the hanky, the automated desktop diary, and the software agent empowered to purchase. In reply to such an observation the skeptic is likely to invoke some kind of ultimate authority: "*Who* was it that decided to knot the hanky," the skeptic demands? "The biological brain, that's who, and that's YOU. *Who* was it that empowered the software agents to purchase? The same old brain, the same old YOU!"

This reply, however, is a major hostage to fortune. Suppose we now ask some parallel questions *within* the neurobiological nexus itself. Do we now conclude that the real "me" is to be identified only with those select elements of the neural machinery involved in *ultimate* decision making? Suppose only my frontal lobes have the final say? Does that shrink the physical machinery of mind and self to *just* the frontal lobes? What if, as the philosopher Daniel Dennett suspects, no neural subsystem has always and everywhere the final say? *Has the mind and self simply disappeared?*

What we really need to reject, I suggest, is the seductive idea that all these various neural and nonneural tools need a kind of privileged user. Instead, it is just tools all the way down. Some of those tools are indeed more closely implicated in our conscious awareness of the world than others. But those elements, taken on their own, would fall embarrassingly short of reconstituting any recognizable version of a human mind or an individual person. Some elements, likewise, are more important to our *sense* of self and identity than other.[32] Some elements play larger roles in control and decision making than others. But this divide, like the ones before it, tends to crosscut the inner and the outer, the biological and the nonbiological. Different neural circuits provide different capacities, and all contribute in different ways to our sense of self, of where we are, of what we can do, and to decision making and choice. External, nonbiological elements provide still further capacities and contribute in additional ways to our sense of who we are, where we are, what we can do, and to decision making and choice. No single tool among this complex kit is intrinsically thoughtful, ultimately in control, or the "seat of the self." We, meaning we human individuals, just *are* these shifting coalitions of tools. We are "soft-selves," continuously open to change and driven to leak through the confines of skin and skull, annexing more and more nonbiological elements as aspects of the machinery of mind itself.

Tools-R-Us. But we are prone, it seems, to a particularly dangerous kind of cognitive illusion. Because our best efforts at watching our own minds in action reveal only the conscious flow of ideas and decisions, we mistakenly identify *ourselves* with the stream of conscious awareness. Then, when in our more scientific moments we begin to inquire into the material and physical underpinnings of the mind and self, it can quickly seem as if much but not all of the brain and all the rest of the body, not to mention the surrounding social and technological webs, are just tools for that conscious user. This is the mistake that once led Avicenna, a Persian philosopher, scientist, and physician who lived from 980 to 1037 A.D., to write of his own arms and limbs:

> These bodily members are, as it were, no more than garments; which, because they have been attached to us for a long time, we think are us, or parts of us [and] the cause of this is the long period of adherence: we are accustomed to remove clothes and to throw them down, which we are entirely unaccustomed to do with our bodily members.[33]

But garments for what? To pursue this route is to embrace a hideously disfigured image of the mind and self, privileging a vanishingly small piece of the true personal and cognitive pie.

Our Worlds, Ourselves

A better bet is the vision of the machinery of mind and self powerfully championed by the philosopher Daniel Dennett. Dennett's work sets out to oppose the persuasive image of the Cartesian Theater: the mythical place inside our brains where sensory inputs, thought, and ideas are all inspected by a "central meaner" whose well-informed choices determine our deliberate actions. Dennett marshals a plethora of philosophical, psychological, and neuroscientific evidence against such a view. His target is often thought to be simply the idea of a neural or functional center of *consciousness*. But Dennett's deeper quarry is precisely the idea of a central self, a small-but-potent internal *user* relative to whom all the rest—be it neural, bodily, or technological—is merely a toolkit. Where we hallucinate a central self, some spiritual or neural point wherein our special individual essence resides, Dennett finds only a grab bag of tools and an ongoing narrative: a story we, as the ensemble of tools, spin to make sense of our actions, proclivities, and projects.[34]

I shall not rehearse Dennett's arguments here (though I have done so at some length elsewhere).[35] Instead, I simply note that our technology-based reflections lead us to the very same conclusions.[36] There is *no self*, if by self we mean some central cognitive essence that makes me who and what I am. In its place there is just the "soft self": a rough-and-tumble, control-sharing coalition of processes—some neural, some bodily, some technological—and an ongoing drive to tell a story, to paint a picture in which "I" am the central player.[37]

Imagine a pile of sand, deposited roughly on the ground, which is slowly settling into a stable arrangement of grains. Were the pile of sand self-aware, it too might hallucinate a kind of inner essence—a special grain or set of grains whose deliberate actions sculpt the rest into a stable arrangement. But there is no such essence. The sandpile simply self-organized into a more-or-less stable coalition of grains. Similarly, certain coalitions of biological and nonbiological problem-solving elements (grab bags of mind tools) prove more stable and enduring than others. These configurations have a tendency to preserve and even repeat themselves. When viewed by a conscious, narrative-spinning element, this all looks like the work of some central organizer: the real self, the real mind, the real source of the observed order. Thus is born the image of the self as a crucial yet vanishingly slim slice of the overall problem-solving ensemble (brain, body, cognitive technologies), a slice so slim and elusive that our own neural circuits (my hippocampus, my frontal lobes) can quickly seem like its tools!

The notion of a real, central, yet wafer-thin self is a profound mistake. It is a mistake that blinds us to our real nature and leads us to radically undervalue and misconceive the roles of context, culture, environment, and technology in the *constitution* of individual human persons. To face up to our true nature (soft selves, distributed decentralized coalitions) is to recognize the inextricable intimacy of self, mind, and world.

This is a confrontation long overdue, and it is one with implications for our science, morals, education, law, and social policy; for these are the governing institutions within which we—the soft selves, the palpitating biotechnological hybrids—must solve our problems, build our lives, and cherish our loves. Yet these governing institutions are slow to change. Just as the law lags visibly behind the complex realities of electronic commerce, so too our social structures and value systems lag visibly behind the accelerating cycles of biotechnological interdependence and interpenetration.

It is at this point that any emphasis on new or near-future technologies, exciting though it is, can prove counterproductive. It can prove counterproductive because it invites knee-jerk/denial and resistance rather than constructive critical embrace. If our technological worlds are threatening to leak into our minds and selves, some would say, it is time to seal the exits, batten down the hatches, and foil the invading digital enemy. My guiding idea, that we are *natural-born* cyborgs, is of course an attempt to preempt precisely this species of response. No point battening down those hatches; the fluids are already mingling and have been at least since the dawn of text, and probably since the dawn of spoken human language. This mingling is the truest expression of our distinctive character as a species.

One incident that helps illustrate this came to me by way of a chance encounter with Carolyn Baum, head of occupational therapy at the Washington University School of Medicine in St. Louis, Missouri. The encounter was chance since, although we worked at the same institution, our disciplinary orbits looked far apart. But disciplinary orbits notwithstanding, Carolyn also turned out to be my neighbor in a wonderful turn-of-the-century town house near the university. So we chatted: we chatted in our garden, in the laundry room, on the stairs. When I explained my ongoing work, Carolyn immediately warmed to the theme. Exactly this lesson, she felt, was emerging from her own work with a subpopulation of inner-city Alzheimer's sufferers in St. Louis. These patients were a puzzle because although they still lived alone, successfully, in the city, they really *should not have been able to do so*. On standard psychological tests they performed rather dismally. They should have been unable to cope with the demands of daily life. What was going on?

A sequence of visits to their home environments provided the answer. These home environments, it transpired, were wonderfully calibrated to support and scaffold these biological brains. The homes were stuffed full of cognitive props, tools, and aids. Examples included message centers where they stored notes about what to do and when; photos of family and friends complete with indications of names and relationships; labels and pictures on doors; "memory books" to record new events, meetings, and plans; and "open-storage" strategies in which crucial items (pots, pans, checkbooks) are always kept in plain view, not locked away in drawers.[38]

Before you allow this image of intensive scaffolding to simply confirm your opinion of these patients as hopelessly cognitively compromised, try to imagine a world in which *normal* human brains are somewhat Alzheimic. Imagine that in this world we had gradually evolved a society in which the kinds of scaffolding found in the St. Louis home environments were the norm. And then reflect that, in a certain sense, this is *exactly* what we have done. Our own pens, paper, notebooks, diaries, and alarm clocks complement *our* brute biological profiles in much the same kind of way. Yet we never say of the artist, or poet, or scientist, "Oh, poor soul—she is not really responsible for that painting/theory/poem; for don't you see how she had to rely on pen, paper, and sketches to offset the inadequacies of her own brain?"

Taking soft selfhood seriously invites us to reconsider our views and prejudices concerning cognitive rehabilitation and the understanding and depiction of cognitive impairment. The forceful relocation of a home functioning Alzheimer's patient into a controlled hospital setting often constitutes a tragic turning point. Such relocation can be akin to the infliction of new brain damage upon an already compromised host. As a society, however, we do not yet enjoy a structure of laws and social policies that recognizes this deep intimacy of agents and their cognitive scaffoldings. The moral is: certain harms to the environment are simultaneously harms to the person. *Our worlds, ourselves.*

All this matters, yet it is easily missed. It is especially easily missed given the recent explosion of interest in evolutionary psychology. Evolutionary psychology treats the bulk of human cognition as a set of adaptations to the specific requirements of a Pleistocene hunter-gatherer lifestyle.[39] Extreme versions of evolutionary psychology depict our minds as hunter-gatherer minds, which subsequently acquired a kind of veneer of technology and tools. There is truth in this, but there is danger also. The truth is that many of our cognitive biases are indeed products of our evolutionary past; the danger is seeing these biases as determining and delimiting the potential of the modern mind. To do this is to misconceive our own brains, which were designed by nature to be unusually open to profound reconfiguration by the specific and technologically evolving environments in which they grow and learn. It is also to ignore, or deliberately downplay, the crucial fact that any built-in neural adaptations are simply one contribution to the developmental unfolding of a complex distributed cognitive device. That complex device is the human mind, and it is a device whose problem-solving routines are defined over an unruly mass of biological and nonbiological circuits and pathways. This is not to reject evolutionary psychology so much as to invite the careful consideration of many more layers of interactive complexity. It is to locate any fixed genetic resources as one small group of players on a crowded stage. Our self-image as a species should not be that of ancient biological minds in colorful young technological clothes. Instead, ours are chameleon minds, factory-primed to merge with what they find and with what they themselves create.[40]

In suggesting that our best biotechnological unions may deeply impact our narrative sense of self, I mean to be suggesting nothing more—but nothing less—radical than the kinds of changes we have already encountered several times in human history. The advent of personal timekeeping made possible a new kind of attitude to life for the average worker. By allowing us to budget our time and divide it between various tasks, we became able to pursue a wider variety of projects. The use of text allowed us to undertake massive intellectual projects, requiring slow, step-by-step critical appraisal and re-appraisal. The selves we construct reflect the specific patterns of opportunity that our cultural, physical, and technological environments provide.

Clynes and Kline, originators of the very term "cyborg," were, we saw, somewhat opposed to the idea that new developments might lead to the bioelectronic *transformation* of humanity into something post-human. Instead, the cyborg additions would simply allow the control of certain bodily processes to fade into the background, freeing the mind to pursue its own, paradigmatically human ends. What all this neglects, however, is the powerful sense in which our conceptions of ourselves (of who, what, and where we are) depend, at several levels, upon the specifics of just this backdrop. My sense of my own physical body depends on my experiences of direct control, and these can be extended, via new technologies, to incorporate both new biomechanical attachments and spatially disconnected, thought-controlled equipment. My sense of myself as the protagonist in my own ongoing story is conditioned by my understanding of my own capacities and potentials—an understanding that must surely be impacted, in deep and abiding ways, by the technological cocoons in which my projects are conceived, incubated, and matured. Such extensions should not be thought of as rendering us in any way post-human; not because they are not deeply transformative but because we humans are naturally designed to be the subjects of just such repeated transformations!

The promised, or perhaps threatened, transition to a world of wired humans and semi-intelligent gadgets is just one more move in an ancient game. It is a move, however, that provides a wonderful opportunity to think longer and harder about what it *should* mean to be human. It helps dramatize the condition we have been in all along and holds up a useful mirror to our current, largely unwired selves. We are already masters at incorporating nonbiological stuff and structure deep into our physical and cognitive routines. To appreciate this is to cease to believe in any post-human future and to resist the temptation to define *ourselves* in brutal opposition to the very worlds in which so many of us now live, love, and work.

CHAPTER 6

Global Swarming

Slugs, Ants, and Amazon.com

A mild winter's morning in Norbury, South London. The sun is freshly risen, and there is coffee steaming in the pot. I look into my mother's backyard and it is awash with glistening, sticky signatures. The walls and paving stones gleam and sparkle with narrow, silvery undulating trails: unmistakable evidence of the nocturnal passage of common garden slugs.

There was a time, now dimly recalled, when I found the mucal scribblings of these small creatures less than entrancing. The trails were, I felt, merely the unsightly by-products of an eccentric mode of locomotion. Today I am enthralled. These glimmering trails are not mere ambulatory by-products but active elements in a distributed, multifunctional, activity-enhancing grid—key players in a smart world for slugs. These trails record, reveal, and simultaneously help structure slug activity. Our own *electronic* trails, laid down as we access data, buy online, and move physically through a world of intercommunicating information appliances, will likewise play multiple important roles in shaping our collective cyborg destiny. By exploiting these trails we will automatically generate new knowledge as we read, buy, and act. It's a global electronic free lunch, and the appetizers are already on the table. But first, the slugs.

The mucus trail laid by an ambulatory slug is made largely of water, with some salts and glycoproteins. It can last up to forty days in some species.

The glycoprotein molecules are the source of all its distinctive sticky properties. The production of this complex chemical goo clearly represents a major metabolic investment for these small beings. In fact, it appears that making the mucus costs slugs and snails about ten times as much as the outlay (in energetic terms) of other animals on running or swimming. Seventy percent of the energy generated by food consumption goes into mucus production.[1] A good question to ask then is, Is it worth it? Is the trail just a rather wasteful side effect of an extremely inefficient mode of locomotion, or is there more to this mucal monorail than meets the eye? Several recent studies suggest the latter: the slug's trail, it seems, is pregnant with unexpected functionality.

One function of the goo (the obvious one) is to facilitate individual slug locomotion. The slug literally glides along the expensive chemical pathway, but the pathway isn't essential and slugs can still move without it, albeit more slowly and with greater difficulty. A second, perhaps less obvious role, is to allow other slugs to follow the same pathway with a much more minimal production of goo. The trails thus function a bit like common highways, smoothing the way for many travelers. A third, and still less obvious function, is to reveal to subsequent travelers the direction of travel of the last slug to use the highway. Chemical traces in the slime preserve direction-specific information that can be "read" by the next slugs along. Finally, and perhaps most unexpectedly of all, the trails are not just roadways but active spawning grounds (courtesy of their high nitrogen content) for the algae on which certain slugs feed—gourmet slug food as a free roadside attraction. Major trails cut collective locomotion costs, convey potentially useful information to new travelers, and perform a kind of farming function to boot.[2] Those ubiquitous silver slimeways are thus multifunctional, activity-enhancing artifacts whose production clearly merits their significant metabolic costs.

The pheromone trails laid by foraging ants provide another, perhaps better-known, example of the use of trails. Consider the foraging behavior of the Argentine ant, *Linepithema humile*.[3] These ants, while seeking food, lay down a distinctive pheromone trail. Now imagine that there are two food sources, one nearer the nest than the other, and that randomly exploring ants discover each source. The ants returning from the closer food source follow their own trail, which now becomes marked with *twice the* concentration of pheromone. The same applies, of course, to the ants returning from the more distant source. But the ants whose total route out and in is the shortest arrive back first, and the pheromone concentration on that trail is therefore temporarily greater. So new ants set out on that trail and, on return, again increase the amount of pheromone, causing even more ants to choose that trail. This process of "positive feedback" (in which successful foraging leading to increased pheromone concentrations, which leads to still more successful foraging, leading to yet another increase in pheromone concentration . . .) allows the colony to rapidly self-organize in order to discover and exploit the best, meaning shortest) routes before gradually moving on—once the nearby food is exhausted—to the next closest source, and so on.

But what, you may well ask, does all that have to with us. A few perfumes and pheromones aside, we humans seem noticeably lacking in native trail-laying skills. Here the contemporary cyborg has a distinct edge, for she is already an electronically tagged agent, swimming in an unremarked sea of intercommunicating information appliances. As we move in physical space and act in commercial and intellectual space, we can automatically lay electronic trails, which can be tracked, analyzed, and agglomerated with those laid by others. Already, trails laid down during web-based search, purchasing, and communications can and are used to inform and personalize the relations between buyers and vendors. Take the mundane business of buying a CD from a firm such as amazon.com. Suppose, to be concrete, you are about to purchase the latest Nick Cave CD. You are told, on-screen, that "people who bought this CD also bought . . . " You are then presented with a list of other CDs purchased by other purchasers of works by Nick Cave. This apparently pedestrian little trick is, in fact, astoundingly potent. Many times I have been led, via the purchasing paths laid down by others who share some aspect of my tastes or interest, to find new and wonderful things, well suited to my somewhat peculiar tastes. To appreciate the full value and potential scope of such techniques we must first understand a little more about how they work, placing them in the larger context of what are sometimes called "self-organizing knowledge structures."

The CD-suggesting technique used by Amazon depends upon a technique known as "collaborative filtering." I first learned about this while visiting the Complex Systems Modeling Team at Los Alamos National Laboratory in

New Mexico. Luis Rocha, a member of the team there, introduced me to a way of thinking about such techniques as exploiting the basic principles of "swarm intelligence." In swarm intelligence, relatively dumb individual agents (ants, bees) create beautiful, complex, and life-enhancing structures (foraging trails, honeycombs, hives) by following a few simple rules and by automatically pooling their knowledge courtesy of chemical traces and structural alterations laid down by their own activity.

Collaborative filtering, as Norman Johnson (head of the aptly named Symbiotic Intelligence Project at Los Alamos) notes, exploits very similar principles to those underlying pheromone-based self-organization.[4] Each episode of use or access by a consumer lays down a trace, and after a sufficient amount of consumer activity, exploitable patterns emerge. Suppose, then, each purchaser of the Nick Cave CD also knew of, and purchased, two other CDs. One is a gift for a friend whose tastes are very different indeed; the other is a CD the purchaser thinks she might like for herself. The chances of substantial overlap in consumer choices concerning the nongift CD are much greater than in the case of the gift. So the trails that get doubly and triply marked by this self-selected group (buyers of Nick Cave's latest CD) are indeed more likely to lead to products that will appeal to the rest of the group (the Cave-lovers). The simple tactic of allowing consumer activity to lay down cumulative trails thus supports a kind of *automatic pooling of knowledge and expertise.*

One reason this kind of procedure is so potent is because it allows patterns of consumer action to speak for themselves and to lay down tracks and trails in consumer space as a by-product of the primary activity, which is online shopping. Those collective tracks and trails have the great advantage of sidestepping all the simplistic categories that we human beings use to classify our own choices. For example, instead of classifying a Nick Cave fan as belonging to this or that category and *then* offering suggestions based on that act of pigeonholing, the "category" is created on-the-hoof by the consumer activity of many Nick Cave fans. If many Nick Cave fans were also listening to The Handsome Family or Peaches, then despite the lack of any obvious common category, these will indeed be duly suggested. Notice how deeply and genuinely different this is from a traditional system that simply assigns each CD to a category (i.e., Patsy Cline = C&W) and then offers you top sellers from that category. "Categorization" by cumulative trail laying is unplanned, emergent, and as flexible as consumer choice itself. Later in this chapter we will see how the same kinds of consideration can be applied to the development of internet search engines, so as to sidestep the rigidity of a the typical keyword-based approach.

Providing the electronic environments that best support flexible, unplanned, collectively self-organized modes of information extraction, retrieval, and organization is immensely important if we are to press maximal benefit from the burgeoning web of human knowledge. I recently purchased a copy of Neil Gershenfeld's excellent treatment of the near-technological future, *Things That Think*. I was appalled to see on the back cover the simple categorization "Non-Fiction: Computing." Such a classification leaves out at least half and probably much more of the ideal readership, which includes artists, designers, engineers, philosophers, and cognitive anthropologists. My initial reaction, however, was surely inappropriate because we increasingly live in a world in which the rather arbitrary labeling decisions—made by well-intentioned vendors—can be trumped by emerging and self-reinforcing patterns of consumer choice. If just a few artists and designers discover and purchase the book, their purchasing trails (which may combine the purchase with others more traditionally suited to their disciplines) will bring the book to the attention of others in their group. All this will happen automatically, as a result of the trails laid during consumer activity. The vendors need never know. The traditional labels need never alter. Now, instead of being consigned to one dusty corner of a physically organized bookshop, Gershenfeld's book lives at the complex intersection of multiple purchasing trails and is equally and simultaneously "present" in multiple viewing locations. Moreover, these locations will shift and alter over time in response to an open-ended set of continuously constructed purchasing trails. The books themselves are, in a sense, actively tracking their best contemporary audiences!

We are merely scratching at the surface. Returning to the original CD-buying scenario, imagine a slightly more sophisticated system, which still exploits past *combinations* of consumer choice. You buy Nick Cave and Patsy Cline, and the system offers you suggestions based on the buying habits of those other folks who are fans of *both* artists. Given a substantial history of consumer trail laying, such a system should automatically track the buried commonality that binds Cline and Cave into a coherent whole.

Finally, imagine a system that retains a trace not just of *what* different individuals purchased but of the temporal *sequence* in which they did so. Such a system might fluidly track common patterns of taste-evolution, and thus offer useful hints about what to try *next*.

Perhaps CD buying is not your bag. How about phone-call or internet message routing? In these cases, messages move from A to B via some series of intermediate stops or switching stations. New routing techniques being studied by France Télécom, BT (British Telecommunications), and MCI Worldcom all employ trail-laying techniques to great advantage.[5] Imagine that each call, as it passes through an intermediate link, lays down a trail. Uncongested links, allowing the rapid free flow of multiple calls, will quickly accumulate an attractive "scent." Now suppose that the traces evaporate over time. The scent deposited in a blocked or slow and congested area will soon disappear, and the link will become unattractive. Some preset degree of random exploration can allow "dead" links to become gradually open again as a few calls pass successfully through. Potent variants include having each call adjust its "scent deposit" according to how long it took to pass through the link, and so on. Such a system continuously self-organizes into an efficient overall message-passing configuration, without any central authority or global monitoring system. Grassroots computation at its finest.

Better Living Through Search

This general idea, of strengthening and weakening connections and trails as an automatic result of ongoing patterns of use, may one day turn the world wide web itself into a kind of swarm intelligence. Another Los Alamos-based group, the Distributed Knowledge Systems Project, has pioneered a kind of self-organizing web server called the Principia Cybernetica Web.[6] The key feature of the server is its ability to create, enhance, and disable links between pages as an automatic result of use. More popular links become increasingly prominently displayed, instigating the kind of positive feedback process described earlier, while little used links wither and fade away. The server can also create new links using a technique that one of the system's originators, Francis Heyligher of the Free University of Brussels, calls "transitivity." Roughly, if many users move from a site A to a site

B and then on to C, it will instigate a direct link from A to C as a kind of shortcut. Returning full circle to the theme of individual human-machine mergers, servers may one day do all this on something more like a user-by-user basis. When you log on, you will be recognized and the hyperlink structure partially adapted to suit you.

To begin to grasp just how very different this would be, consider what a similar degree of user-sensitivity would look like were it (impossibly) realized in the physical world of roads and interstates. Our roads, too, are nonbiological structures, which alter and transform our needs and projects; they are largely fixed and static structures, slow to alter and respond to changing needs and pretty well impervious to the quirks of individual road users. Imagine then a world in which the roads and interstates automatically adapt and change, re-routing themselves in response to patterns of use. Little used routes become smaller and then fade away, busy routes automatically expand, varying according to the time of day, and there is automatic re-routing around congested areas. Most spectacularly of all, the whole road network slightly reconfigures itself in response to your personal tastes each time you step into your car. A *Traffica Cybernetica* would be something like that!

The best of the new generation of internet search engines, although they do not actively restructure hyperlink space, nonetheless work by exploiting the collectively created knowledge implicit in the links between web pages. They mine the knowledge implicit in the multiple trails (in this case, the hyperlinks between web pages) that structure the collectively created web. First generation search engines such as AltaVista and Infoseek relied heavily on fairly simple forms of first-order heuristic search, ranking pages according to the number of times the query items appeared, or how early in the text they did so. Such engines often retrieved, even when used properly, voluminous junk and had a regrettable tendency to miss the good stuff altogether. There is a sense in which this is not surprising, for the problem they confront is formidable. There are often literally millions of pages whose contents look superficially relevant, especially given that the usual test for relevance is dumb syntactic matching: the search engine seeks pages that either contain, or are indexed as containing, tokens of the specific string or strings entered by the user. The situation is worsened by the unplanned, anarchic nature of the web itself: there is little deliberate global organization of the kind that might be useful in streamlining search. Second-generation search engines, such as Google, have found an interesting way around this problem. The key trick, it seems, is to focus attention not (ultimately) on the content of the pages so much as on the *structure of links between pages*. The hyperlink structure itself—the way different pages link to and from each other—turns out to be a treasure house of communally generated implicit knowledge concerning which pages are most central and authoritative regarding a given topic. In 2001 the Cornell University computer scientist Jon Kleinberg received the National Academy of Science award for Initiatives in Research for his work on such methods. Kleinberg devised a set of algorithms or formal methods to extract and utilize some of the knowledge implicit in the burgeoning web of connectivity.[7] To illustrate the scale of the problem, Kleinberg takes the sample query string "Harvard." It so happens that

> there are over a million pages on the WWW that use the term "Harvard" and www.harvard.edu is not the one that uses the term most often, or most prominently, or in any way that would favor it under a text-based ranking function. Indeed, one suspects there is no purely *endogenous* [internal] measure of the page that would allow one to properly assess its authority.[8]

Here are a few other examples: to search for "search engines" is especially tough because many of the most authoritative pages (Yahoo, Excite, AltaVista) do not use that term on their home pages; to search for very broad topics, such as "censorship," tends to return a hodgepodge of largely nonauthoritative sources. Standard searches thus tend to be both inefficient (returning too much) and insufficiently intelligent (despite returning too much, they often miss—or return way down in the list—the most relevant and authoritative sites).

Kleinberg's procedure *starts*, nonetheless, with a dumb text-based search. It collects a number of the pages returned for some broad query by a standard search engine. This delivers a "root set" (R) of pages: a set that, as just noted, is quite likely to *fail* to contain the pages in which you are, in fact, most interested. The next step is to seek a set of pages that is more likely to contain the pages you need. The key assumption here is that the authoritative pages, though perhaps missing from the root set, are often at least *linked to* one or more of the pages in that set. The set is thus expanded to include all the pages directly hyperlinked to pages in the root set, along with some key restrictions to keep things manageable.[9] The next step is to consider the number of pages in this new set that link to, and from, other pages in the set. This reveals which pages are, in a sense, the most popular in this new set. The trouble is that mere popularity doth not authority make. Indeed, some sites are almost universally linked-to, regardless of topic: amazon.com is a prominent example. Additional filters are clearly required. Kleinberg notes that if a site is indeed authoritative with respect to a query, we may expect not only that many other sites in the expanded set link to it but that there will be certain pages or "hub pages" that have links to several such authorities. The best authorities, likewise, will be linked to multiple such "hub pages." The heart of Kleinberg's procedure is therefore an algorithm that computes, using second-order information about *patterns in hyperlink space*, this mutually interdefined set of hubs and authorities. Let us pause to appreciate the results. Here, for example, are the top search results for the query strings "java," "censorship," "search engines," and "Gates." The numbers on the left indicate the overall strength of the "authoritativeness" rating:

(java) Authorities

.328 http://www.gamelan.com/
Gamelan

.251 http://java.sun.com/
JavaSoft Home Page

.190 http://www.digitalfocus.com/digitalfocus/faq/howdoi/html
The Java Developer:How Do I . . .

.190 http://lightyear.ncsa.uiuc.edu/;slsrp/java/javabooks.html
The Java Book Pages

.183 http://sunsite.unc.edu/javafaq/javafaq.html
comp.lang.java FAQ

(censorship) Authorities

.378 http://www.eff.org/
EFFweb—the Electronic Frontier Foundation

.344 http://www.eff.org/blueribbon.html
The Blue Ribbon Campaign for Online Free Speech

.238 http://www.cdt.org/
The Center for Democracy and Technology

.235 http://www.vtw.org/
Voters Telecommunications Watch

.218 http://www.aclu.org/
ACLU: American Civil Liberties Union

(search engines) Authorities

.346 http://www.yahoo.com/
Yahoo!

.291 http://www.excite.com/
Excite

.239 http://www.mckinley.com/
Welcome to Magellan!

.231 http://www.lycos.com/
Lycos Home Page

.231 http://www.altavista.digital.com/
AltaVista: Main Page

(Gates) Authorities

.643 http://www.roadahead.com/
Bill Gates: The Road Ahead

.458 http://www.microsoft.com/
Welcome to Microsoft

.440 http://www.microsoft.com/corpinfo/bill-g.htm[10]

To compare this to dumb text-based search, it is useful to note that almost *none* of these pages appeared in the root set R (the original set of pages returned by text-based search). They instead first appeared in the expanded set obtained by adding the pages linking in and out, and obtained *prominence* only by the further computation of likely hubs and authorities. In fact, the *only page* in the above list that was returned by the original text-based search was www.roadahead.com, returned as the 123rd choice of AltaVista!

There is something wonderfully reassuring about all this. First of all we created, by a mass of anarchic, individual efforts, a global web of information whose main drawback is that *because* it is so large and generated so haphazardly, there is no central index and no effective methods for finding what you need, when you need it. Instead of trying (hopelessly, I feel) to remedy this by central planning (e.g., by artificially imposing rigid structure and order, perhaps restricting the ability of unauthorized individuals to create and post information, creating ever-more complex systems of coding and indexing) we now find that *anarchy is its own best medicine*. Our distributed, centrally uncoordinated efforts already encode, in the electronic spaghetti of hyperlink trails between pages, a mass of hard-won, collectively generated knowledge about which sites are most important. That knowledge can now be rapidly and flexibly accessed using tools (such as Google and Kleinberg's somewhat fancier, but related, procedure) that focus more upon the abstract structure of the collective hyperlink weave than upon the contents of specific pages.

Second-order, hyperlink pattern-based search is thus a potent tool for accessing and deploying those vast knowledge bases that we are collectively creating. Ease, speed, and accuracy of access will be the crucial determinant of the social and psychological impact of new knowledge-based technologies. If our sense of the limits and extent of our own knowledge is to be altered and expanded by the use of such tools, the tools need to find and deliver the right stuff, at the right time, reliably and with a minimum of user effort. Powerful wearable computers, with wireless communications links and well-tailored, invisible-in-use interfaces are an important step forward. Lacking intelligent, flexible, adaptive search and retrieval routines, this is just the gateway without the goodies. Both are of the essence.

It is interesting to see, to take one familiar example, how deeply Google has already altered our relations to the web. Better search engines make the extensive use of electronic bookmarks redundant. It is now simpler and quicker to enter a modestly well-chosen search string than to hunt through a giant file of prestored bookmarks. In 2001 the science and technology magazine, *New Scientist*, announced that for this reason it was abandoning its "Netropolitan" column, a listing of web sites relevant to hot topics, on the grounds that a simple search using a good engine will generate an up-to-the-minute listing on the spot.[11]

The presence of large, accessible, easy-to-search web-based resources is poised to impact our research and publishing activities in profound ways. James O'Donnell is a professor of classical studies whose speciality is the life and work of St. Augustine. He is also provost for information systems and computing at the University of Pennsylvania and has thought long and hard about books, computing, and the potential synergies that combinations of new and old technologies may offer for academic studies. One important effect, O'Donnell suggests (and it is one that our discussion of collaborative filters and new search routines already anticipates) will be on the organization, storage, and dissemination of academic materials. Instead of needing to decide in advance (as we do with physical storage) where to place a copy of a text, links to the text can be placed in many locations in hyperspace; the text can be found, given good search techniques, by just about anyone who cares about it—as long as he has access to such technologies.

Powerful search routines allow a bunch of relevant materials to be recruited, grouped, and organized on-the-hoof into a kind of "soft-assembled" information package. Soft assembly is a useful concept, which I first encountered in the work of the developmental psychologists Esther Thelen and Linda Smith. Smith and Thelen describe the way multicomponent systems can sometimes self-organize in order to exploit a useful subset of elements or resources, creating a temporary stable structure that solves some adaptive problem.[12] Think, for example, of the human capacity to walk. Walking is soft assembled insofar as it naturally and pretty much automatically adjusts, in detail, to accommodate new conditions. Icy sidewalks, high-heeled shoes, a blister, and a sprained ankle all recruit different patterns of gait and muscle control while aiming at the common goal of efficient locomotion. The child's limbs grow bigger and stronger almost daily, but she does not need to relearn how to reach, eat, and play tennis. One of the keys to such successful soft assembly is a kind of "equal partners" approach in which bodily, environmental, and neural factors all cooperate without any central overseeing control to solve the problem, in whatever way it is currently and perhaps temporarily presenting itself.[13]

A soft assembled information package, in this sense, is a package whose elements are not tied together firmly by fixed, preexisting links. Instead, they are brought together on the spot in response to a specific query, made by a specific user, in a specific context. Contrast this with the kinds of information packages with which we are still most familiar: a book, a single edition of a journal, or a set of pamphlets assembled by a firm or company. In all these other cases, what you get is what someone else, at some previous moment, and sometimes for reasons of economy or availability rather than content, decided to stitch together.

Luis Rocha (who we met earlier in this chapter) is an expert on Distributed Information Systems, which he defines as "collections of electronic, networked resources in some kind of interaction with communities of users."[14] The internet, the web, corporate intranets, and databases are all examples of such systems. The more such systems can be set up to be self-organizing, changing, and evolving in automatic response to changing patterns in user activity, the closer we come, Rocha suggests, to a kind of collective human-machine symbiosis. Early information retrieval routines, based solely on keyword searches and the like, were limiting and inflexible tools. Their demerits were compounded by the fact that the databases they were required to search were often large, unruly, and with no central organization or standard format. These retrieval routines were passive and rigid, unable to lead the researcher in new, appropriate but unanticipated directions—unlike even the simple collaborative filtering techniques deployed by amazon.com. They were all-purpose, making no attempt to tailor their searches to a long-term profile of the user; and they suffered from what Rocha dubs "fixed semantics," having no way to amend and update their own search and indexing as a community evolves new terms, ceases to use older ones, and begins to link together previously unrelated areas of study.

Active Recommendation Systems, based on the kinds of collaborative filtering techniques described earlier, address all these problems. Such systems are sometimes called Knowledge Mining Systems. A good example is TALKMINE, developed by Rocha as a test-bed application for the research library at the Los Alamos National Laboratory. TALKMINE deploys a software agent (just a piece of active code) to retrieve, select, and filter documents, and to conduct an initial question-and-answer routine used to establish a rough profile of an individual user's needs and interests. This information is used not simply to guide the current search but also to amend and update the long-term memory bank of associations that constitute the bulk of TALKMINE's "intelligence." For example, if several users start to combine keywords that were not previously closely associated in the system's memory, it will create a new association that can then be used to guide future suggestions to other users. The long-term memory of such associations is the system's main tool for selecting and filtering information from the databases to which it is linked, but it is a tool that is automatically and continuously changing in response to user input, and which alters to accommodate changes in the terminology and expectations of a community of users. It is not bounded by any fixed semantics, and by using collaborative filtering techniques combined with keyword strategies it can actively point the user in new and useful directions. It thus establishes what Rocha describes as "an open-ended human-machine symbiosis . . . facilitating the rapid dissemination of relevant information and the discovery of new knowledge."

Bundling information into preset, pretagged physical packages (such as books, journals, and so on) may thus become less and less crucial, as users learn to soft assemble resources pretty much at will, tailored to their own specific needs. One result of this may be the gradual erosion of the firewalls that currently separate various documents and sources of information.

> Preservation of the boundaries separating one piece of information from another is necessary only if the end-user needs it, and search strategies that concentrate on the information rather than the source are far more efficient and will thus be more successful.[15]

O'Donnell also notes that often he doesn't care about the source, just about the information (as when we consult a dictionary or an encyclopedia, for example). It is worth noting, though, that we all care about *reliability*, even if the precise *source* of some piece of information is unimportant to us.

To a certain extent, search engines that deploy versions of Kleinberg's procedure are able to serve both these needs at once. They target information rather than sources, but they are sensitive to the collective judgments of the community of users regarding which sources are most authoritative and give these higher priority. There are, of course, often good reasons to care about sources. A paper published in a major journal, for example, has passed severe peer-based tests before making it into print. Electronic media will probably need to preserve some version of this; but we should not, in so doing, attempt to directly "police" or control publication on the web. What O'Donnell proposes instead is the separation of the idea of *validation* from the idea of *prepackaging*. Print journals conflate the two and validate submissions by publishing them in preset packages. In the electronic world, major journals might instead add (after the usual kinds of peer-reviewing process) a kind of seal of approval to certain articles. A single article could carry the seals of multiple major journals, encouraging consumption by a wider audience. In my own field of cognitive science, I am often dismayed to have to choose between publishing a certain paper in, say, a philosophy journal, versus an artificial intelligence journal. The same article, published electronically, could easily carry the seals of approval of both.

Such proposals, clearly, raise a host of social, legal, political, and economic questions, some of which we will confront in the final chapter of this book.[16] But the point, for now, is simply to flag the increasing viability of on-the-spot soft assembly as a means of accessing and grouping information and resources. In this new arena, the interactions between primary and secondary materials (e.g., original discussions versus commentaries and critiques) will also mutate, as hyperlinked assemblies allow scholars to move directly between different translations, editions, experiments, critiques, and so forth. Important works and results will be the nodes around which whole new kinds of communal effort can congregate and self-organize. New advances will be very much the work of multiple geographically distributed hands, self organized around these central seeds. Digital media, fluent global communications networks, and potent search engines will combine to support global intellectual swarming, creating a common arena in which the pooling, combination, selection, recombination, and mutation of ideas can occur faster and more efficiently than ever before.[17]

Real books do, of course, have certain advantages over electronic documents and even over fresh printouts. A well-used text, as Norman and others have pointed out, bears useful traces of previous patterns of access and use. *Wired* magazine's executive editor, Kevin Kelly, reports Will Hill's ongoing attempts to create a kind of useful virtual analogue to physical wear and tear. A much-consulted physical document (in a shared library, for instance) provides useful information in the brute evidence of wear and tear. Well-thumbed pages are probably crucial, or they are especially problematic—worth pausing over. Hill, a Bellcore researcher, aims to enrich electronic documents in a similar way. A spreadsheet might reveal, by color-coded highlighting, which figures had been most often or most recently revised. A similar trick allows a programmer to see where, in a few million lines of C++, the most recent changes have been made. Some of these functions are already commercially available. In Hill's own lab, as Kevin Kelly notes, electronic documents go a few steps farther, keeping detailed records of their own patterns of use:

> When you select a text file to read, a thin graph on your screen displays little tick marks indicating the cumulative amount of time others have spent reading this part. You can see at a glance the few places other readers lingered over. . . . Community usage can also be indicated by gradually increasing the type size. The effect is similar to an enlarged "pull quote" in a magazine, except these highlighted "used" sections emerge out of an uncontrolled collective appreciation.[18]

Once more, automatic electronic trail-laying provides the aromatic whiff of an informational free lunch; but like every free lunch, this one too has its dangers. Once we come to rely on these highlighted passages, we may actively ignore the rest of the document. In doing so, we ourselves add to the emerging stress on the highlighted passages, and we may then confront a kind of self-fulfilling prophecy. Early bad highlighting (due to a few episodes of unintelligent early use) could lead others to linger there, causing more highlighting, causing still more people to linger there, and so on. The danger is thus of a kind of *dysfunctional communal narrowing of attention*, exacerbated by a process of runaway positive feedback or a bad case of what is sometimes also known as early path-dependency.

We shall return to these kinds of worry in chapter 7. For now, I simply note that one solution is simple awareness, on the part of the users, that this is how it works and that these are the dangers. Such users will take the highlighting seriously but always with a pinch of salt. Even a cursory glance at the rest of the document might help avert major errors, and nudge the communal enterprise back on course.

Starlogo, SimCity, and the Global Informational Free Lunch

How do we train young minds to think better about swarm-like systems? Brains like ours are not ideally suited to the task. But we can now build designer learning environments tailored to instill and support better habits of thought. Our biological brains, in concert with these new technologies, can thus grow into hybrid minds better able to understand the kinds of systems in which they themselves participate.

One such tool is StarLogo, developed by Michel Resnick of the MIT Media Lab.[19] An educational software package aimed to encourage better thinking about decentralized systems, StarLogo allows the child to influence the on-screen behavior of a teeming multitude of simple agents. The original Logo software, probably familiar to many readers, used a few on-screen "turtles" as drawing aids for the programmed creation of geometric shapes and pictures. StarLogo, by contrast, offers whole hordes of mini-agents, each one able to sense its (simulated) environment, to respond according to programmable rules, and to alter the environment as it does so. By varying simple rules governing such responses, the child learns how complex effects emerge from the interactions between many agents and environmental structures.

Such tools can play a role both in early learning and later in adult life, by supporting the detailed simulations that inform frontline research. Human minds will thus grapple ever more successfully with the kind of decentralized complexity that characterizes so many critical systems, from highways to ant colonies to the world wide web to human minds themselves. Earlier we suggested a simple moral: *know thyself, know thy technologies*. New educational technologies such as StarLogo take this one step farther. They *are* technologies, which can help us better to know our soft, decentralized selves.

StarLogo is, however, limited in one crucial respect. It offers superb support for thinking about simple swarm-like systems comprised of multiple entities all or most of which are obeying the same simple rules. Our biotechnological self-image, however, depicts the human individual as a swarm-like ecology with multiple *heterogeneous* parts. Enter SimCity.

Created by Will Wright in the early nineties, SimCity is an interactive video game that has sold well over a million copies. The goal, as the name suggests, is to create and maintain a virtual city—but you don't get to do this by any kind of micromanaging design activity. Instead, you must nudge, sculpt, and tweak your city using various forms of indirect control. SimCity teaches co-control for complex, heterogeneous assemblies. The city may have a mayor, but the mayor does not get to micromanage either. Think of the mayor as just one more source of tweaks and nudges to the complex system: by gently manipulating a few variables, such as zoning and land prices, you may be able to bring about some effect, for example, to encourage the building of a new shopping mall. The domino effects here will surprise you, as new ghettos and high crime areas emerge in its wake. The bigger the city, the more complex the interactions. The skill of "growing" a thriving, happy city is precisely the skill of embracing co-control. It is the skill of *respecting the flow*, while subtly encouraging the stream in some desired direction.[20]

StarLogo, SimCity, and its recent companion "The Sims" are designer environments that can help biological brains learn to get to grips with decentralized emergent order. They can help us develop skills for understanding those peculiar kinds of complex systems of which we ourselves are one striking instance. Experience with such tools should be compulsory elements in our educational practice.

Returning to that worry about the progressive narrowing of attention via techniques such as collaborative highlighting, another kind of solution beyond simple awareness is more technology. Certain users might, for example, be able to make a bigger difference to the electronic trails than others. The time these users spend on passages might lay down a triple trail, for example, and this, in turn, might depend upon how well the community had rated that individual's previous trail-laying over time. Such multilayered setups require further infrastructure, but the potential rewards are high, as they enable us to create communities that automatically police and regulate their own processes of automatic knowledge creation.

It is also worth noting that to reap the benefits of global intellectual swarming need not ultimately mean abandoning the paper-based book in favor of a nasty, flickering, unreliable computer screen in a hot dusty corner. The simple, paper-based book, as Neil Gershenfeld nicely notes, provides information in a robust, portable, easily annotatable, easy-to-view form.[21] Fluorescent tube–based electronic books are still a clumsy halfway house, often displaying the worst, rather than the best, of both worlds. But maybe—just maybe—we can one day have it all. A team based at MIT Media Lab has been finding ways to make real paper behave like a computer display, adding new functionality while preserving the best of the old interface and format.[22]

Real paper reflects light in specific ways that make the pages easy to read and is tolerant of multiple viewing angles and distances. The amount of power needed to illuminate a printed page is tiny when compared with the power needed to drive a fluorescent tube-based display. The drawback is that the paper display is fixed and noninteractive, isolated from other agents and the global database. This may all change. The Media Lab team is experimenting with paper covered with microencapsulated particles. Carbonless copy paper already works like this, using tiny ink-filled shells that are selectively broken by the pressure of the pen on the top sheet. In the hi-tech version, the shells contain particles—some black, some white. These particles have different electrical charges and can be manipulated in an electrical field. The result is a kind of toner-like electronic ink. Paper thus treated can be printed and then directly recycled, since all the printer does is to rearrange these black and white particles. Add electrodes sandwiched inside each sheet, and the paper itself can do the job. Add a small solar cell and a tiny radio receiver printed on the page, and all you need do is leave the pages in the light to allow them to retrieve and display new information as required.

Such technologies, Gershenfeld cautions, are not yet fully viable, but some of the main ideas and techniques are in place. The point to stress is that as we contemplate (and as we design and commission) new technologies, we should not be too quick to assume that acquiring distinctive new functionality means sacrificing the old. Paper is robust, portable, easy-to-read, and (once printed) power-efficient. Electronic media, communal trail-laying, and intercommunications provide for rapid updates, free data-mining, and personalized interactive services. We can, and should, aspire to both. Above all, we should beware of visiting the sins of the interface on the deeper underlying technologies themselves. Our interfaces to the electronic world are still, for the most part, nasty, clumsy, fragile, and slow—but things are changing fast.

One rather familiar way in which fast, fluent, global information sharing already supports new kinds of collaborative creation is via the use of freeware and open source code for software development and testing. The operating system Linux is a case in point. In August 2001, some ten years after its creation, Linux controlled 27 percent of the world server market and was the dominant alternative to the Windows operating system of Bill Gates.[23]

Developed by Linus Torvalds of Finland in 1991, Linux was licensed under open source rules (GPL or general public license) that allow anyone to download the code for free and to write and distribute amendments. When companies, such as Red Hat, act as distributors for such open source properties, they cannot vary this arrangement. Instead, they make their profit from packaging, support, and services. The downside is that they must pay their own programmers to write improvements, which are then given away. The upside is that they can reap the benefits of a global community of users who improve, test, and debug the software for free. The amazing thing is that no corporation owns Linux, and no legally demarcated group of individuals is responsible for its reliability, operation, or upkeep. It is, instead, a distributed labor of global love—a kind of grassroots movement mounted against the domination of Microsoft.

A *New York Times* article in 1998 quotes Randy Kessel, a manager at Southwestern Bell who, almost as a dare, installed Red Hat Linux on thirty-six desktop PCs that monitor operations in Kansas and Missouri. Mr. Kessel notes the characteristic combination of virtues and vices. The virtues are cost, reliability, and the rapid-fire free support of an international community of users. The vices are long-term reliability (i.e., what if the community grows tired of supporting the product?), privacy, and accountability. As he puts it:

> We took a mission-critical operation and we deployed a free operating system there . . . and now we spend a tenth of the administrative costs for these desktops that we do for the rest of the 315 that we use [but] the legal department says "when it fails, who do we sue?" The I.T. department says "it is not a proved product." Corporate security says "it's hackerware." But it's the only thing that worked.[24]

More on those vices later. The virtues are clear. By swarming, pooling, and sharing our knowledge, as well as by exploiting the empowering context of systems for collaborative filtering and automatic data-mining, we are slowly creating that Global Informational Free Lunch. And the goodies do not stop there. Human agents will not be the only ones sending and sharing information through these new networks. Search engines and software agents are already out there, working the web on their own, bidding, buying, selling, and searching. Next on the scene will be our old friends, the Information Appliances. Here, for example, is Neil Gershenfeld on the future of health care:

> Billions of dollars are spent annually just taking care of people who didn't take their medicine, or took too much, or took the wrong kind. In a TTT [Things That Think] world, the medicine cabinet could monitor the medicine consumption, the toilet could perform routine chemical analyses, both could be connected to the doctor to report aberrations, and to the pharmacy to order refills, delivered by FedEx (along with the milk ordered by the refrigerator and the washing machine's request for more soap). By making this kind of monitoring routine, health care could be delivered as it is needed at a lower cost, and fewer people would need to be supervised in nursing homes, once again making progress on a hard problem by building interconnected systems of simple elements.[25]

Of course, the amount of new net traffic that will be generated when everyone's toilet, fridge, medicine cabinet, or wet bar, has online capability will be large. This will require precisely the kinds of cheap, effective, automatic message-routing strategies described earlier. In attempting to support these large distributed systems of intercommunicating parts, the use of self-organization and electronic swarm-based solutions will be crucial indeed.

Some proponents of Ubiquitous Computing and Information Appliances believe that as our worlds become smarter and more self-organizing, our own personal needs to access information will grow less and less. Of this I am skeptical. Such a development is neither likely nor desirable. Our thirst for knowledge and entertainment, and our drive to understand and create, will not go away just because our fridges know when to order more milk. The day is unlikely to come when people no longer feel the need to privately access and use the web. We will, however, live in a world increasingly well suited to the laying of ever more complex electronic trails. We ourselves will be electronically tagged by various forms of simple wearable computing, and we will be moving among a dense backdrop of signal-sensitive, intercommunicating devices. A world populated by Information Appliances among which electronically enhanced people move and work is a world ripe for trailing and tracking. A simple visit to the zoo or the shopping mall could soon be guided by collaboratively filtered suggestions: "People who visited such and such locations, and bought such and such goods, also liked these locations. . . . "

The possibilities are as liberating as they are simultaneously worrying. If each person bears a distinct electronic signature, then individual movements, preferences, activities, and histories can be automatically recorded and agglomerated. Issues of privacy and security loom large, but these we have postponed until the next chapter. For a moment, at least, let's dwell on the good stuff. Imagine, again following Neil Gershenfeld, a world in which your particular driving habits can—if you allow it—be monitored and analyzed by the insurance company, courtesy of electronic links with your cars.[26] Good drivers whose cars are well-maintained may then get cheaper rates than others. If you choose not to share such information, you will simply not get the discount and the policy will be priced the old-fashioned way. Norwich Union, a major UK-based insurer, announced in early 2002 plans to offer reduced premiums based on the installment of GPS (Global Positioning Satellite) technology, which could record information about the car's movements.[27] If you use the car only for short daytime trips, you might get a lower rate than someone who travels long distances at night. Nor is there any need for this system to require your car to let the company know where you are at all times. The system can be set up to share *only* the kinds of information, and at the level of detail, deemed appropriate.

Information concerning your personal driving skills and the type and condition of your car might, Gershenfeld suggests, be used to negotiate personalized speed limits with police monitoring stations. Taking into account your skills, the car you drive, the road, traffic, and weather, you may see a speed limit of 80 mph on the Augmented Reality roadside display, while someone else, driving along the same stretch of road, may see 65. Drivers choosing not to share the required kinds of information would simply remain bound by a standardized limit of the usual 60 mph. Similar kinds of personalized approaches might be available for life insurance and health insurance.

For better or for worse (and almost certainly for both), human-technology symbiosis is poised to transform our lives both as individuals and as collective groups. At the individual level, new transparent technologies will increasingly blur the already fuzzy boundary between the user and her tools for thought; at the collective level distributed activity-sensitive software will enable us to press new knowledge from electronic trails of use and access. These trails of use and access will also support more fluid and personalized services, while better search engines and user-responsive systems will enable us to retrieve, bundle, and deploy information in new, fluid, soft assembled ways. These shiny new tools will not simply redistribute old knowledge; they will transform the ways we think, work, and act, generating new knowledge and new opportunities in ways we can only dimly imagine. Our smart worlds will automatically become smarter and more closely tailored to our individual needs in direct response to our own activities. The challenge, as we are about to see, is to make sure that these smarter worlds are our friends, and that our tracks, tools, and trails enrich rather than betray us.

CHAPTER 7

Bad Borgs?

Needless to say, the grass isn't always greener on the cyborg side of the street. Under the rocks of our new liberties and capacities lurk new closures, dangers, invasions, and constraints. It is time we confronted some of the specters that haunt these hybrid dreams. They include

Inequality	Overload	Deceit
Intrusion	Alienation	Degradation
Uncontrollability	Narrowing	Disembodiment

Let's look them in the eye.

Inequality

Some of us, to be sure, are comfortably cocooned in our biotechnological nests. We will benefit directly from new waves of human-centered technology, and we are happy nodes in those larger swarms of connected consumption, choice, and learning. But is this just another trick to cement an unfair, unequal world order? By the year 2004, on optimistic estimates, at most 10 percent of the world's population will enjoy easy internet access.[1] On the other hand, in 1999, 70 percent of the world's population had never made a phone call.[2]

The wired (and increasingly wireless) world is smaller than we might at first imagine. All this is quite compatible with my claim that a profound openness to deep human-machine symbiosis is part of our basic human nature. But shouldn't we be wary of a world in which this potential is realized by only a very few? Yes, we should—but not by artificially stifling the biotechnological impulse.

Neil Gershenfeld, a director of the M.I.T. Media Lab whom we already met in previous chapters, tells the story of a visit by a group of people from developing countries. The visit was organized by a Media Lab–associated foundation called 2B1, whose brief is to introduce appropriate and useful forms of information technology to children in African villages and in other developing areas. Gershenfeld anticipated deep concerns about "cultural imperialism," but encountered instead an eager and open spirit. The strong opinion of the delegates (admittedly a self-selected group) was that "the world is changing quickly and it is . . . elitist to insist that developing countries progress through all of the stages of the industrial revolution before they're allowed to browse the web."[3]

The major challenge, on this view, overlaps considerably with the kinds of research and development scouted in previous chapters. For new technologies to be a help rather than a hindrance in developing regions, they need to be cheap, robust, and intuitive-to-use—in a word, human-centered. "Such a thing," Gershenfeld wryly comments, "is not in the direct lineage of a desktop PC."[4] But it *is* in the direct lineage of work that seeks to make advanced informational technology fade into the background of our daily lives and projects. These technologies *must* be cheap (so they can be ubiquitous), and they must be robust and intuitive-to-use. Wearable computing, for example (see chapter 2) aims to free the user from the tyranny of the electric grid and, rather than using costly and environmentally harmful battery packs, will eventually exploit the user's own activity (e.g., walking) to generate the power required.

I am not suggesting, not for one moment, that these new robust, human-centered technologies are a panacea for global inequality. Far, far, from it. The inequalities are endemic, they are massive, and they won't go away anytime soon. The question is whether we should fear—in this push toward greater biotechnological symbiosis—a whole new wave of inequalities: the cyber haves versus the cyber have-nots. Here I remain guardedly optimistic. As our best technologies become less fragile, cheaper, and easier-to-master, more doors open to more people than ever before. MIT's recent decision to make most of its undergraduate and graduate course materials available on the web, free of charge, to any user anywhere in the world, is a cause for celebration.[5] In a similar vein, a new initiative from the World Health Organization will give free internet access to a thousand top medical journals to libraries and universities in the world's sixty-five poorest countries.[6]

Taken as a package, the emerging wired (and soon to be wireless!) world is neither intrinsically good nor intrinsically evil. It is simply up to us, in these critical years, to try to guarantee that *human*-centered technology really means what it says: that human means all of us and not just the lucky few.

Intrusion

You live and work in a smart world, where your car is talking to your coffee machine (and snitching to your insurance company), and your medicine cabinet and toilet are watching your inputs and outputs (and snitching to your doctor or HMO, not to mention the drug police). Your smart badge (or maybe your cell phone) ensures that your physical movements leave a tangible trail, and your electronic trail is out there for all to see. The damn telemarketers know your soul; their machines have surfed your deepest likes and dislikes. So whatever happened to your right to a little space, some peace and privacy, a quiet affair, a little psychotropic time-out?

For my own part, I am delighted that Amazon, courtesy of some neat collaborative filtering, is able to recommend some stuff that I really do want to hear. But do I really want the government—or worse, Microsoft—to have access to all my movements, ingestions, consummations, and consumptions? The joys of the electronic trail and ubiquitous computing suddenly pale against the threats of electronic tattling and ubiquitous interference.

A few real-life horror stories can help set the scene. Consider the cookies. Cookies are electronic footprints that allow web sites and advertising networks to monitor our online movements with granular precision. DoubleClick, a major internet advertising company, was able to place cookies on millions of consumer hard drives. As a result, you might find yourself the target of unsolicited ads for products related to those you have most recently surfed. Innocent enough at first, but when DoubleClick aquired Abacus Direct, a huge commercial database of names, addresses, and online buying habits, it was able to stitch the information together to link real names and addresses to the cookie-based information about online use.[7] Under public and governmental pressure the so-called profiling scheme was put on hold. But the potential is there. Amazon once deployed a system that identified the books and items most commonly purchased by people at specific major institutions and corporations, using domain names and ZIP codes to zero in. "People at Charles Schwab tend to like *Memoirs of a Geisha*"—that kind of thing.

Scarier still are the GUID's (Globally Unique Identifiers). These get pinned to you when you register for a service online and allow the company to link your online activity to your real-world details. Similarly, some Microsoft wares embed a unique identifier into each document you create allowing it to be traced back to its real author.[8] It is well known that many companies and corporations, in blatant invasion of reasonable expectations of privacy, monitor e-mail even when it is sent from home over a company network. The trouble, as Jeffrey Rosen (see note 5) nicely points out, is that the more such intrusions occur, and are not legally blocked, the lower our expectations become. The law is set up to protect our privacy in proportion to our reasonable expectations—a nasty little circle if ever you saw one. Reduce your expectations, and your rights follow suit. It is up to us, the public, to make sure that our expectations of privacy are not unreasonably eroded. We must not be browbeaten by disclaimers ("your e-mail may be monitored") whose legality is often quite untested. Correlatively, it is a matter of extreme urgency that the courts proceed with great care when making new law in this area. Privacy, once lost, is often lost forever.

Ubiquitous Computing only compounds the problem. The natural support systems for information appliances and swarm intelligence equally and naturally provide for an unprecedented depth and quality of surveillance. It is one thing for your liquor cabinet to tell the store you need a new bottle of Ardbeg Single Malt; quite another when it tells your employer that you seem to be drinking more than can be good for you. The simple cell phone emits a signal that can act as a "smart badge," talking to all those semi-intelligent appliances you pass on your daily rounds. Now your friends, family, employers, lovers, and even your lovers' lovers, need never be at a loss concerning your current whereabouts. Turn it off or leave it behind?

Once the cell phone apparatus is lightly implanted in the skull, you won't even be able to accidentally leave it behind, though God knows, we'd better be able to turn it off. But then how does THAT look when your boss wants to find you?

Smart-badge systems, which allow the firm to track an employee's on-site movements, have already been tested at XeroxPARC, EuroPARC, and the Olivetti Research Center.[9] Really smart ones, of course, need to know who has just entered and what kinds of things they are likely to want. Worse still (but better for automatic data-mining) they need to remember exactly what previous visitors did, or bought, or accessed. In the era of ubiquitous computing and swarm intelligence, walls really *do* have ears, and memories too.

One response is just to bite the bullet—just do the calculations and decide that, on the whole, there is more good than bad in the creation of a fluid, adaptive, personalized, and rapidly responsive environment. If our health improves, or insurance costs go down, and we are always traceable in case of an emergency, who's to complain? If I am offered goods and services that I actually want, at prices that I am happy to pay, why worry? In life, there are always trade-offs, and if the price of all these benefits is a certain loss of privacy, maybe that's a price we should be prepared to pay. As Scott McNealy, C.E.O. of SUN Microsystems, once famously remarked, "You already have zero privacy: get over it."[10]

Perhaps we can have our cake (in private) and eat it (ubiquitously, in public). To a certain extent, at least, technology itself has the potential to allow us privacy when we choose it. Wearable computing has a role to play here, as does the impressive work on public-key encryption. Wearables, as Bradley Rhodes and his colleagues point out, can help by keeping a lot of data quite literally on the person, instead of distributing it through a variety of intercommunicating fixed-location devices.[11] The trouble, of course, is that if we *don't* allow outside agencies sufficient tracks, trails, and histories, we cannot reap the benefits of recommendation systems, personalized services and pricing, and the like. Likewise, if we don't allow the tool to know who we are, it won't be able to serve us so well. The price of any islands of privacy and disconnection is thus a reduction in the range of support and responses automatically available. Certain goods and services may also cost us more if we are unwilling to share personal information with the providers, but these trade-offs should be ours to choose. What matters most—and this is a lesson we will return to again and again in this closing chapter—is that our technologies be responsive to our individual needs, including our occasional but important need for privacy.

Encryption helps. By using clever (often *so* clever that they are currently illegal) ways of encoding digitally transmitted information, we are already able to allow outside agencies as much or as little content-access to our electronic meanderings as we choose. Public-key encryption allows you to advertise a key—a string of numbers—that anyone can use to encrypt a message to you, but the public key alone is insufficient to decrypt or decode anything. It works only in unison with another such key, the one known only to you. Using these kinds of techniques, it is simple to send secure information across the web. Freeware versions of public-key encryption systems include PGP (Pretty Good Privacy, legal *only* in the United States) and RIPEM, which is public domain software distributed by RSA.[12] Advanced cryptography applications support other useful functions, such as the so-called zero-knowledge proof.[13] This allows a merchant to bill a consumer without the merchant learning who is buying or having access to any details of the person's account.

What all this means in practice is that the user can, if she wishes, selectively opt out of some of the trailing and tracking swarm-based infrastructure. She can buy goods and services without revealing personal information and can block or filter the transmission of identity-revealing data to the various information-appliances surrounding her. Currently, the default is extreme openness to intrusive surveillance, and only the technologically sophisticated tend to take the various steps needed to protect themselves. Such users might, for example, employ advanced security tools such as Kremlin, which combine encryption capabilities with programs that are able to genuinely delete unwanted files from your hard drive.[14] This requires writing nonsense strings on top of the remaining chunks of files that standard programs simply partially delete. It helps avoid the Monica Lewinsky syndrome, where prosecutors subpoenaed her home computer and ransacked the hard drive, retrieving long-deleted and never-sent love letters to the president of the United States.[15]

Such extreme measures can smack of paranoia. As firms and legislators will inevitably argue, why worry unless you have something to hide. But worry we do. Most of us would hate to live in a state where government and employers routinely opened our physical mail. Why should we live in a state where the natural successors to such mail (e-mail and text messaging, phone calls and faxes) lack similar protections? Moreover, the whole notion of "having nothing to hide" is fuzzy to say the least. It covers interests and activities ranging from the outright illegal to the mildly perverse to the simply socially frowned-upon to the merely eccentric. The average upstanding citizen may well have "something to hide" once this broad church is canvassed.

We live, after all, in a society where a great deal of behavior, which is neither illegal nor harmful, might if made public impact negatively upon our lives and careers. Spare a thought for the grade school teacher who likes to cross-dress in the privacy of his own home but buys the clothes off the web, and visits other sites and chat rooms to share discoveries and experiences. Or the peace campaigner with a taste for violent literature and a one-click account with Amazon. Or the Catholic priest with a nuanced love of women's lingerie. Or, to end with a real-life example from the British press, the gay police chief with a soft spot for anarchy and cannabis. I leave the reader to fill in her own special interests. The list is endless, the shades of gray innumerable.

And then there is the e-romance. Anyone foolish enough to attempt to conduct an extramarital, or otherwise unauthorized, affair using electronic media will quickly find cause to regret those tracks, trails, and incomplete deletions. They spell doom for your dreams of a private corner of cyberspace, complete with white picket fence and a full range of modern domestic appliances.

Then there are drugs. In an age when large numbers of well-informed, intelligent adults occasionally partake of recreational drugs (other than the taxed, time-honored, and often quite lethal alcohol), it might be hoped that they will do so in as careful a manner as possible. To that end, they might visit a useful site such as Dancesafe, which offers balanced information concerning doses, effects, addictiveness, and relative toxicity. Yet if such visits are perceived as a two-edged sword, perhaps helping users avoid the worst kinds of abuse but simultaneously adding their names and details to some law enforcer's list, can we really hope for such care and caution?

All this is disturbing since it again hints at the creation of a new elite: this time, the elite subset of internet users who stand any chance of achieving even a modicum of privacy. For every one who deploys advanced security tools such as Kremlin and public key encryption, there will be a thousand who trust to the goodwill of commerce and government. In any case, it seems unlikely that many citizens, be they ever so technologically aware, will ever win an "arms race" between users and government/employers. Encryption and firewalling is probably not the ultimate answer.

Another possibility, which I have grudgingly come to favor, involves a kind of leap of faith, or democratic optimism. As our governments, employers, colleagues, and lovers learn more and more about the typical behavior of a wide range of valued, productive, and caring citizens, it should become clearer and clearer in what ways the goalposts of "good behavior" must be moved. Such movement need not signal decay and decline, for only hypocrisy and more solid public/private firewalls ever kept them in place! As the lives of the populace become more visible, our work-a-day morals and expectations need to change and shift. It is time for the real world to play catch-up with our private lives, loves, and choices. As the realm of the truly private contracts, as I think it must, the public space in any truly democratic country needs to become more liberal and open-hearted. This attitude, which I am calling democratic optimism, may seem naively idealistic, but it is surely preferable to an escalation of cyber wars that the average citizen simply cannot hope to win.

Uncontrollability

Some suggest that we should actively limit our reliance on technological props and aids, not just to protect our privacy but to control our own destinies and preserve our essential humanity. Here, the title of this book gives me away. Human-machine symbiosis, I believe, is simply what comes naturally. It lies on a direct continuum with clothes, cooking ("external, artificial digestion"), bricklaying, and writing. The capacity to creatively distribute labor between biology and the designed environment is the very signature of our species, and it implies no real loss of control on our part. For *who we are* is in large part a function of the webs of surrounding structure in which the conscious mind exercises at best a kind of gentle, indirect control.

Of course, just because nature is pushing us doesn't mean we have to go. There are times, to be sure, when the intelligence of the infrastructures does seem to threaten our own autonomy and to cede too much, too soon, to the worlds we build. In the novel *Super-Cannes*, J. G. Ballard depicts a highly engineered environment ("Eden-Olympia—the first intelligent city") in which

> there are no more moral decisions than there are on a new superhighway. Unless you own a Ferrari, pressing the accelerator is not a moral decision. Ford and Fiat and Toyota have engineered in a sensible response curve. We can rely on their judgment, and that leaves us free to get on with the rest of our lives. We've achieved real freedom, the freedom from morality.[16]

Chilling stuff. The more so since this vision of machines bearing more and more of the load once borne by biological intelligence is *precisely* the one with which Clynes and Kline launched our cyborg odyssey, back in 1960 and back in chapter 1.

We are torn, it seems, between two ways of viewing our own relations to the technologies we create and which surround us. One way fears retreat and diminishment, as our scope for choice and control is progressively eroded. The other anticipates expansion and growth, as we find our capacities to achieve our goals and projects amplified and enhanced in new and unexpected ways. Which vision will prove most accurate depends, to some extent, on the technologies themselves, but it depends also—and crucially—upon a sensitive appreciation of our own nature.

Many feel, for example, that increased human-machine symbiosis directly implies decreasing *control*. In an age of Ubiquitous Computing must we be slaves to the whims of the machines that surround us? In an age of global swarming, should we fear that even the machines don't have a leader? Have we cast ourselves as King Lear, but with whole legions of ungrateful daughters?

As we saw in chapter 5, the kind of control that we, both as individuals and as society, look likely to retain is *precisely the kind we always had: no more, no less*. Effective control is often a matter of well-placed tweaks and nudges, of gentle forces applied to systems with their own rich intrinsic capabilities and dynamics. The fear of "loss of control," as we cede more and more to a supporting web of technological innovations is simply misplaced. What matters is not that we be micromanaging every detail of every operation, but that the surrounding systems provide usable, robust support for the kinds of life and projects we value. This is *precisely* the goal of human-centered technologies anyway. The trick, then, is to acclimatize ourselves to a much more *biological* relationship with our technologies.[17] We should neither expect nor desire to give detailed instructions to our machines; instead, we should be able to simply factor their *capacities* into our own work and projects.

In a strange kind of way, it is the semi-autonomous machines that hold out the best prospect of one day constituting integral parts of distributed, biotechnological, hybrid intelligences. Our conscious intelligence would work *with* these devices in much the same way as conscious vision (recall chapter 4) works *with* the semi-autonomous processes supporting fine visuomotor activity. The complementary skills of these biological subsystems help make human intelligence what it is today. The complementary skills of a host of nonbiological subsystems will help make human intelligence what it is tomorrow.

Overload

One of the most fearsome specters, though far less abstract and dramatic, is that of plain simple overload—the danger of slowly drowning in a sea of contact. As I write, I am painfully aware of the unread messages that will have arrived since I last logged in yesterday evening. By midday there will be around sixty new items, about ten of which will require action. Ten more may be pure junk mail, easy to spot or filter, but it is the rest that are the real problem. These I read, only to discover they require no immediate thought or action. I call this mail e-stodge. It is filling without being necessary or nourishing, and there seems to be more of it every day.

The root cause of e-stodge, Neil Gershenfeld has suggested, is a deep but unnoticed shift in the relative costs, in terms of time and effort, of *generating* messages and of *reading* them.[18] Once upon a time, it cost much more—again in terms of time and effort—to create and send a message than to read one. Now, the situation is reversed. It is terribly easy to forward a whole screed to someone else, or to copy a message to all and sundry, just in case they happen to have an opinion or feel they should have been consulted. The length of the message grows as more and more responses get cheaply incorporated. Other forms of overload abound. The incoming messages aren't all e-mail; there are phone calls (on mobile and land lines) and text messages, even the occasional physical letter. There is the constant availability, via the Google-enhanced web, of more information about just about anything at the click of a mouse.

One cure for overload is, of course, simply to unplug. Several prominent academics have simply decided that "e-nough is e-nough" and have turned off their e-mail for good or else redirected it to assistants who sift, screen, and filter. Donald Knuth, a computer scientist who took this very step, quotes the novelist Umberto Eco, "I have reached an age where my main purpose is not to receive messages." Knuth himself asserts that "I have been a happy man ever since January 1, 1990, when I no longer had an e-mail address."[19]

We won't all be able to unplug or to avail ourselves of intelligent secretarial filters. A better solution, the one championed by Neil Gershenfeld, is to combine intelligent filtering software (to weed out junk mail) with a new kind of business etiquette. What we need is an etiquette that reflects the new cost/benefit ratio according to which the receiver is usually paying the heaviest price in the exchange. That means sparse messages, sent only when action is likely to be required and sent only to those who really need to know—a 007 principle for communication in a densely interconnected world. E-mail only what is absolutely necessary, keep it short, and send it to as few people as possible.

Alienation

Warwick University campus, in the English city of Coventry, was the relaxed and convivial setting for the Fourth International Conference on Cognitive Technology. The meeting took place in August 2001, and I was blessed with the task of delivering an opening keynote address. Adopting my usual, upbeat approach, I spoke of a near future in which human-centered technologies progressively blur the already fuzzy boundaries between thinking systems and their tools for thought. I addressed problems and pitfalls, but mostly of a technical or methodological kind. What I hadn't anticipated—especially from this well-informed, enthusiastic crowd of scientists—was the amount of real ambivalence that many felt toward a future in which so many of our interactions would be with so-called agent technologies instead of with flesh and blood humans. This, in fact, turned out to be a major discussion topic throughout the conference.

One version of this fear was articulated by John Pickering of the Warwick University Psychology Department. In a wonderful talk peppered with memorable (if sometimes disturbing) images[20] from the media, Pickering painted a worrying picture. Agent technologies, he suggested, may "harmfully degrade how people value themselves and treat each other."[21] By "agent technologies" he had in mind the kinds of long-running, potentially interactive, software packages discussed in chapters 1 and 6. Examples might include a web-searching agent who seeks out the kinds of antique books and records you desire, reporting back, and bidding on your behalf; a "chat-bot" that you call up when you are feeling lonely or depressed; or a semi-intelligent interface that allows you to tell your graphics program, in plain English, what you seek to achieve and then discusses your plans in the light of its own deeper knowledge of what the underlying software can and can't do.

This kind of application is especially important because the kinds of biotechnological merger and symbiosis we have been discussing may well depend, for their ultimate success, on the creation of just such biofriendly interfaces: software agents that know enough about human language and human psychology to grease the wheels of human-machine interaction. Surrounded by a host of such agents, from a very early age, Pickering fears for the child's basic understanding of what it is to be human. For these "technologized social interactions" may well never be as deep, sensitive, and caring as the best of our interactions with other human beings. The software agents may mimic aspects of our social interactions, but they will do so (for the foreseeable future at least) only shallowly and imperfectly. The worry is that by exposure to such mimicry, our *own* view of ourselves and others may become warped, altered, or downgraded, and "the process of sociocultural learning in which human identity is formed will be changed as a result."[22]

The kinds of agents that Pickering is most concerned about are the learning agents: the software entities that adapt to you, learning about your likes, dislikes, tolerance for detail, best times for contact, and so on. Such agents, he thinks, will be perceived as individuals and will impact our ideas about our own "spheres of responsibility." A child who has a pet dog is already interacting with a simple form of intelligence, but a good software agent will be able to mimic more advanced aspects of human social interaction. It is this, Pickering worries, that might lead them to treat real human beings as more like software agents—to value both equally, and perhaps as a result to "dumb down" the human-to-human interactions they engage in. Kirstie Bellman of the Aerospace Corporation likened this process to the adoption of a "spell checker vocabulary" when sending messages to one another. If we know a word the spell checker doesn't but we aren't sure of the spelling, we tend not to use it. In this way, our active vocabulary for human-to-human interaction gets dragged down to the level of what the spell checker knows! Imagine, then, a scenario in which a child, interacting with a software agent that understands only a few simple emotional expressions, actually ends up limiting even her interactions with her parents to that same level of mutual understanding. This is a terrifying prospect indeed.

Pickering has a point. We really do need to pay closer attention to the many ways in which new technologies may impact our social relations, and our sense of ourselves and of others. As identity becomes fluid, embodiment multiple, and presence negotiable, it is the perfect time to take a new look at who, what, and where we are. New kinds of human-machine symbiosis will, without a doubt, alter the way we see ourselves, our machines, and the world. As N. Katherine Hayles, a University of California professor, rather eloquently puts it, "When the body is integrated into a Cybernetic circuit, modification of the circuit will necessarily modify consciousness as well. Connected by multiple feedback loops to the objects it designs, the mind is also an object of design."[23]

Our redesigned minds will be distinguished by a better and more sensitive understanding of the self, of control, of the importance of the body, and of the systemic tentacles that bind brain, body, and technology into a single adaptive unit. This potential, I believe, far, far outweighs the attendant threats of desensitization, overload, and confusion. A few comments, though, on that specific worry about children's (and adult's) use of software agents.

My own reaction, at the conference, was to present a kind of benign dilemma. Either the software agents would be good enough to really engage our social skills, or they wouldn't. If the former (unlikely), then why worry? But if the latter (much more likely), then the child would still engage with her human caregivers in a visibly different way. Just as having a pet tortoise does not make a child less likely to want to play catch with her parents, having highly limited interactions with software agents won't blind her to the much wider range of interactions available with her parents.

A different kind of response came from Kirstie Bellman herself. All these worries, she suggested, actually rang *less* true to her than to her male colleagues. For Bellman is a busy working scientist who is also a mother. Even if interactions with software agents and play-bots are somewhat shallow, Bellman argued, they can act as a useful supplement to the richer social interactions that are (all sides agreed) so crucially important. Busy working parents simply need all the help they can get. Just as previous generations of children loved, cared for, and talked to their dolls and pets, so new generations might add software entities to this venerable list. The kinds of fears and worries that so exercised many (predominantly male) members of the group were, she felt, a kind of luxury item freely available only to a certain professional class. This is an important point. Critics within the scientific community, such as the psychologist John Pickering of Warwick University in the UK, often fear that "enthusiasm for . . . computer-enhanced lives usually comes from a highly visible and technologically sophisticated minority [and that this] tends to conceal the more negative experiences and views of the less technologically adept majority."[24]

Bellman turns this on its head, pointing out that it is equally often only the lucky few who have the luxury of fearing the effects of labor-saving and opportunity-enhancing innovations such as microwaves, dishwashers, and, one day, software agents who play with the kids. What about the already real interactive toy that is designed to help children learn to share? The toy is a doll-filled castle that the child shares with a 3D computer-animated playmate whose image is projected onto a screen of the castle. The playmate tells the child stories about the dolls. But if the child makes a premature grab for a doll, the virtual playmate politely objects, reminding her that he was still playing with it. The playmate also helps structure cooperative play between many children by telling appropriate stories. Early studies suggest that children using the interactive toy learn to behave better, as a result, with their real playmates.[25] Perhaps, then, there is hope for new interactive toys that help, rather than hinder, a child's social and moral development. In the end, what really matters is that we educate ourselves and our children about *the nature and the limits of our best technologies*, so that we can intelligently combine the best of the biological and engineered worlds.

The Cognitive Technology conference was also the occasion for a wonderful encounter with Steve Talbott of the Nature Institute. Steve is probably the best informed and most constructive critic of the role of advanced technologies I have ever met. He and I disagree about just about everything, but there is no better guide to the dangers of alienation inherent in the biotechnological matrix. I strongly recommend taking a look at Steve's monthly electronic publication NETFUTURE (subtitle: *Technology and Human Responsibility*). At the time of writing, NETFUTURE is well into its second hundred issues, and it covers everything from the complex issues surrounding technology for the handicapped ("Can Technology Make the Handicapped Whole?" NETFUTURE #92) to the role of computers in childhood ("Fools Gold," in #111) to the question of e-mail overload and the balance between connection and disconnection (#124). In a typically nice twist, Steve argues that technologies that enable distant communication (e-mail, cell phones, etc.) are a double-edged sword. For while they can help bring us closer together, the also create the *conditions* under which more and more of us are *willing* or required (by our firms) to move physically farther and farther apart. The simple presence of these technologies thus contributes to the generation of the very problem (frequent, easy, long-distance communication) they help to "solve." In the end, Steve's point is *not* that we should therefore give up on cell phones and e-mail accounts. Instead, he says:

> Our failure to recognize the truth about the technological forces we are dealing with . . . prevents us from bending them more effectively to our own ends. If we came to terms with the double significance of our technologies . . . we would not so routinely speak of cell phones, e-mail and the like in terms of the single virtue of connectivity. We would recognize that the underlying forces of disconnection at work in these tools are fully as powerful as the forces bringing us together.[26]

By keeping a weather eye on the darker side of our technologies and inviting us all to participate in the discussion, NETFUTURE performs an invaluable and ever more timely service. Tune in at http://www.oreilly.com/people/staff/stevet/netfuture or just plug "netfuture" into a good search engine.

Narrowing

Consider the simple use of a software agent to suggest new books for you to read or new music for you to hear. Such an agent will make its recommendations on the basis of (*a*) its knowledge of what books or CDs you have bought before, (*b*) your feedback, if any, concerning which books or CDs you liked best, and (*c*) its knowledge—courtesy of the collaborative filtering and data-mining techniques discussed in chapter 6—of what books or CDs others, who liked the ones you liked, liked too.

This is all well and good, as far as it goes, but Patti Maes, of the MIT Media lab, argues that there is an attendant danger of a kind of communal tunnel vision. Such software agents will suggest more and more of what are broadly speaking "the same kinds of thing," to the same kinds of people. Choosing from these lists, these people will then confirm the software agent's "expectations" by buying (and probably even liking) many of these things. So the agents will, in effect, offer us more and more of a progressively less and less extensive band of literature, music, or whatever. The danger is of a kind of positive-feedback-driven "lock-in" following a few (perhaps ill-chosen) early purchases or decisions.

Compare this with a visit to a bookstore, where a bright jacket or a snappy title might catch your eye, and where your trip to the detective fiction section takes you right past poetry and cooking. The real world, it seems, currently offers a much richer canvas for semirandom explorations than does its virtual counterpart. But a real-world bookshop, as we all know, is often less than ideal when you already know exactly what you want; stocks are limited, organization unfathomable, opening times idiosyncratic. The potential synergy between real-world browsing and online targeted purchasing is truly enormous. A few lucky discoveries can seed whole new areas of interest, which (once reflected in your online purchasing or even earlier if the cash registers talk to your software agents) will add new dimensions to your electronic profile, and hence give the software agents and collaborative filters lots more avenues to explore.

The moral is simple but just about maximally important. To really make the most of the wired world, we need to understand—at least approximately—how it works. Only then can we take the measure of its weaknesses and its strengths, and adjust our own role, as human participants, accordingly. Technological education will be crucial if human-machine cooperation is to enrich and humanize rather than restrict and alienate. Once again, the lesson seems clear: *Know Thyself: Know Thy Technologies.*

Deceit

In 1996 there appeared an article in *Emerge* magazine called "Trashing the Information Highway: White Supremacy Goes Hi-Tech."[27] It revealed the increasing use of the internet as a means of conducting devious smear campaigns. In one such campaign white supremacists posing as African-Americans posted offensive calls for the legalization of pedophilia. This is a pernicious abuse of the ease and anonymity of the internet. Such abuses are fortunately uncommon and fall more or less under the authority of existing law.

More common and far less easy to control (but much harder to classify as abuse rather than innocent self-reinvention and exploration) is the use of electronic media to present a sexual or personal persona that is in some way different from the sender's biological or real-world persona. Chat rooms are full of (biological) men presenting as women, (biological) women presenting as men, (biological) older women presenting as younger women and vice versa, (biological) younger men presenting as older men and vice versa, gay (biological) men presenting as straight women, straight (biological) men presenting as gay men. The permutations seem endless, a kind of Goldberg Variations on all the rich sexual, social, and physical complexity the nonvirtual world has to offer.[28] In one chat room someone recently admitted to presenting herself as a multiple amputee even though she was biologically quite intact.

What should we make of all this? On the one hand, it seems like deceit: an impression reinforced if we view the electronic domain as a kind of hunting ground for possible real-world encounters. Those who use the media this way often resort to quick and dirty early checks, like an impromptu request for an immediate telephone conversation. But clearly, nothing is ever conclusive. Instead, the only real hope is that if it is clear that the parties concerned might want to use the net as a springboard to a real-world (and I use that term grudgingly) relationship, then they (and you) will not waste too much time presenting in ways that cannot, with a little goodwill on both sides, successfully carry over into that other context.

But what of the many folk who do not seek to cross the divide? Or those who might one day segue into a real-world meeting but believe that the real "them" is precisely the combination of a certain set of personas, some adapted to the conditions and constraints of their biological form, life situation, and previous history, and others adapted to the very different conditions, constraints, and possibilities presented by these new forms of communication, contact, embodiment, and presence? Might these not be genuine, complex individuals in their own right? Who are we to insist that the real "you" is defined by some specific subset of your words, actions, and interests?

Some of the deeper issues here concern the successful integration of multiple personas, where by "integration" we can mean something quite subtle. To be integrated, in this sense, is not to have one constant persona, so much as to be able to balance the needs of various personas so as to avoid compromising any one by the actions of the "others." This is, in effect, a recipe for distilling a multidimensional form of personal identity from a flux of potentially competing ways of presenting oneself to others and to the world.[29] This might mean, for example, being wary of the strategy of building impermeable firewalls between your electronic and real-world selves, and instead allowing communication, overlap, and seepage. It might mean being able to be honest about your biological self and real-world situation, without taking that as devaluing these other forms of personal growth and exploration. Such an approach will become increasingly practical as more people appreciate the potential of new media to support entirely new forms of personal contact, presence, and relationship, rather than seeing them merely as imperfect attempts to recreate real-world relationships and presence at a distance.

More disturbing, in many ways, is the presence in many chat rooms of nonhuman intelligences pretending to be human. The web portal Yahoo, in 2001, was "infested by cyber-bots."[30] These were programs able to log on to the chat rooms, and posing as humans, send messages directly to other people in the chat rooms, enticing them to visit specific company web sites. Free advertising, with that important personal touch. Cyber-bots can likewise pass themselves off as voters in online polls, or as participants in contests, or in other ways.

To prevent such abuses, researchers at Carnegie-Mellon's Aladdin Center have created what they nicely dub CAPTCHA—Completely Automated

Public Turing-Test to Tell Computers and Humans Apart.[31] The test requires those seeking an account for entry to some space (say a chat room) to take a simple test to "prove" they are human.[32] A word is shown against a background that adds noise or as a distorted version of itself (see fig. 7.1). The prospective account holder must identify the word. This task,

Fig. 7.1 CAPTCHA (Completely Automated Public Turing-Test to Tell Computers and Humans Apart). CAPTCHA aims to unmask web-bots posing as humans by asking them to recognize words and shapes against a backdrop of noise. Illustration by Christine Clark.

simple as it sounds, currently weeds out the bots from the boys (or girls), as present-day word recognition routines cannot cope with these deviant presentations. Once again, it is simply an arms race between competing technologies, and CAPTCHA may not serve as gatekeeper for very long.

Such potential for deceit or dissimulation is balanced, however, by the very real power of new communications regimes to spread important truths quickly, without the usual impediments of censorship and bureaucracy, and without regard for many of the physical, national, and social boundaries that render so much of our daily news parochial in the extreme.

A case in point is the simple e-mail sent by Tamin Ansary to some twenty friends and colleagues on the morning of September 12, 2001. Ansary was an Afghan-American living in San Francisco. The letter, which I am willing to bet nearly every reader of this book received within two or three days of the attacks on the Twin Towers, argued that any U.S. response that involved "bombing Afghanistan back to the stone age" would be misplaced. It would be misplaced not because the crimes were not heinous but because "that's been done. The Soviets took care of it already. Make the Afghans suffer? They're already suffering. Level their houses? Done. Turn their schools into piles of rubble? Done."

This message didn't need testing by gatekeepers to check its authenticity. It smelled of truth the way a diner smells of doughnuts. Those who received it saw this at once, sending it on to their own friends and colleagues. It found its way within a few days to the web sites Tompaine.com and Salon.com, and from that stopover conquered the world.[33] By the end of the week, the message had reached the hearts of millions upon millions of people across the globe. Ansary himself was besieged with requests for interviews, with e-mails, phone calls, and offers.

The message didn't stop the eventual bombing, but it may well have played a role in delaying, and perhaps partially reconfiguring, what was perhaps politically inevitable. At the very least, it gave millions of people new insight into the complex political and social realities upon which simplistic talk of evil and retribution is all too easily overlaid. It stands as a testimony to the power of the internet to allow words and ideas to reach a massive audience, not because those words come with the standard trappings of authority or because they enjoy the brute force backing of standard international media, but simply in virtue of their timeliness and content.

Deceit, misinformation, truth, exploration, and personal reinvention: the internet provides for them all. As always, it is up to us, as scientists and as citizens, to guard against the worst and to create the culture and conditions most likely to favor the best.

Degradation

Close cousin to these worries about deceit is a worry about lack of quality control. In the wired (and soon to be wireless) world, where anyone can publish thoughts and insinuate e-mails into thousands upon thousands of inboxes, how are we to separate the wheat from the chaff? The problem is especially pressing given the very real problem of overload, mentioned earlier. Time is a precious resource, and we cannot afford to read everything everyone has to offer us in order to decide—even assuming we could tell—what is most authoritative or important.

Sometimes, of course, an item might arrive in our inbox with the validation of a close and trusted friend. In the case of the Ansary letter just described, my first copy arrived early in the chain, and with the endorsement of just such a friend. But what happens when materials arrive via a public bulletin board or an unpoliced newsgroup? The alternatives at first seem stark: either we regress to some kind of good old-fashioned central authority (such as reading only the online *Times*), or we confront an unsorted, unfiltered barrage of information, misinformation, and innocent but time-consuming spam. Hope, however, springs eternal, and our choices may not be so stark after all.

Consider the case of Slashdot, a bulletin board serving, at first, a small group of friends in the small town of Holland, Michigan.[34] To start with, the board worked well. The friends shared many interests (Star Wars, video games—you get the picture) and posted only things that most of the group wanted to see. But as time went on, traffic increased, much of it from far away. The board's originator, Rob Malda, was unable to filter the postings. His first response was to appoint some lieutenants—people he trusted to help sieve the spam. Apart from locking out the truly offensive or totally irrelevant, these lieutenants had an added power: the power to rate the remaining contributions on a scale of 1 to 5 (5 being the best). Users of the board could then choose what quality level they wanted to inspect, locking out whatever fell below their chosen level of tolerance. The gradings also served to encourage good postings, since everyone wanted the five-star ratings for their own work. But the population continued to explode. As Steven Johnson reports:

> It was the kind of thing that could only have happened on the web. A twenty-two-year-old college senior, living with a couple of buddies in a low-rent house—affectionately dubbed Geek House One—in a nondescript Michigan town, creates an intimate on-line space for his friends to discuss their shared obsessions, and within a year fifty thousand people each day are angling for a piece of the action.[35]

What could be done? Rather than try the normal remedies, Malda made the entire group collectively responsible for its own quality control. The system worked like this. After a few visits as a Slashdot user, you might find a message telling you that you had been temporarily assigned the role of moderator. At any given time, a shifting subset of users would have this status (rather like being called to jury duty) and would be asked to rank other users' contributions on the 1 to 5 scale. Each moderator is allowed only so many points, and once they are awarded, the moderator ceases her role. On top of this, however, Malda introduced a system he called Karma. A specific user would accrue Karma according to how many of the person's past postings had achieved high ratings. Those with "good karma" got special rewards! New postings from these users would begin with a higher default rating than the others, and the users would more likely be chosen as moderators. The moderation process thus collectively helps choose the moderators themselves. As a result, those whose postings were most highly ranked by the group tended to become the key figures in guiding the group ahead. Best of all, the new system works. A new user can just set the quality control to 4 or 5 and find thousands of recent postings reduced to a few dozen high-quality items, while the more adventurous, or time-liberal, user can still explore the peripheral spaces in search of missed gems.

This broad approach, in which users rate the activities (including especially the rating or reviewing activities) of other users, offers the best current hope for a kind of collective, flexible, grassroots approach to the tricky questions of policing, filtering, and regulating. At its best it preserves most of the delicious freedom and anarchy of the web, while allowing individual users to reduce their cognitive loads and home in on reliable sources more or less at will. Amazon, eBay, and other large web-based concerns have all implemented their own versions of these so-called meta-feedback systems (ones using feedback about the usefulness and quality of feedback) in the last few years.

Finally, as Steven Johnson notes, there is no need to fear that such systems must tend toward narrowing and conservatism (for example, favoring postings that are liked by the average user). Instead, the underlying algorithm could be altered to favor moderators whose choices have sparked the most feedback, or whose own postings have generated large numbers of responses both pro and con, and so on. These moderators would still hunt the spam, but with the overall system thus tweaked, the level 5 filter would now favor not the safe median but the stuff most likely to generate intense debate and feedback. In fact, a single system could easily offer both, allowing users to choose which kind of filter (median or controversial) they prefer. The possibilities thus exist for an open-ended variety of new and potent forms of swarm intelligence, with meta-feedback reconfiguring our filtering routines to suit the different types, or moods, of users.

Disembodiment

I have a special stake in this one, as I have long championed the importance of the body in the sciences of the mind. One of my books even bears the subtitle "Putting Brain, Body and World Together Again." Imagine my horror, then, to find myself suspected, in writing enthusiastically of technologies of telepresence and digital communication, of having changed sides, of now believing that the body didn't matter and the mind was something ethereal and distinct.

Far from having changed sides, however, the present work flows directly from this stress on the importance of body and world. What we have learned is that human biological brains are, in a very fundamental sense, incomplete cognitive systems. They are naturally geared to dovetail themselves, again and again, to a shifting web of surrounding structures, in the body and increasingly in the world. Minds like ours solve problems not by intellectual force majeure but by cooperating with all these other elements in a spaghetti-like matrix. Just about everything in the present treatment speaks in favor of that image, from the use of pen and paper to do complex sums to the ease with which Stelarc now deploys his "third hand," to the daily babble of cell phones and text messages with which we now coordinate our social lives, all the way to the use of mind-controlled cursors, swarm-based data-mining, and telepresence guided house-minding devices. Moreover, as we saw in chapter 2, the intimacy of brain and body is evidenced in the very plasticity of the body-image itself. Our brains care *so much* about the fine details of our embodiment that they are ready and willing to recalibrate those details on the spot, again and again, to accommodate changes (limb growth, limb loss) and extensions (prosthesis, implants, even sports equipment). It is this tendency that allows them sometimes to be fooled by certain tricks, and it is because of this that the physical feeling of remote presence—and even of remote embodiment—is sometimes quite easy to achieve. The brain, in all these cases, is just one player on a crowded field. Our experience of what it is to be human, and our sense of our own capacities for action and problem solving, flows from the profile of the whole team.

Whence, then, the fears about "disembodiment"? One root of the worry is the popular image of the lonely keyboard-tapping adolescent, who prefers video games to human company, takes no interest in sports or direct-contact sex, and who identifies more closely with his or her own electronic avatar or avatars than with his or her biological body. Isolated, disconnected, disembodied, desexed. Virtues, perhaps, in a politician, but hardly what we would wish for any child of our own.

The image itself is open to empirical question. According to a University of Warwick (UK) survey, heavy internet surfers are more likely not less to belong to some real-world community group, and less likely to spend time passively watching TV.[36] Talking to others on the net encourages, it seems, the appreciation that we can get together with like-minded folk and actually make a difference in the world. Nonetheless, the image of the isolated key-tapper is one we seem to have indelibly added to our stock of modern-day stereotypes.

Isolation, in any case, is often a matter of perspective. The apparently isolated individual tapping away night after night is, in many cases, spending quality time in her own chosen community. These eclectic electronic communities often bring together a greater number of like-minded folk than we could ever hope to find in our hometown or even in a large city. A rather bizarre example is the online community of folk who gather at sites such as FurryMUCK.[37] A MUCK is a multiuser (usually role-playing) environment, and FurryMUCK caters to those whose imaginative, social, and sometimes sexual pleasure involves adopting animal personas and/or wearing furry animal costumes. This once-elusive minority now has hundreds of web sites and their own (real-world) conventions and meetings. Without the distance-defying glue of electronic chat rooms and communities, it is hard to imagine such a group achieving this kind of critical mass.

There is, however, a new danger that accompanies the creation of more and more specific (often gated) electronic communities. It is one that is especially marked in the case of communities held together by shared but unusual sexual preferences or tendencies. The danger is of a new kind of marginalization. By relying upon an electronic community in which it is easy to speak of unusual needs and passions, people with special interests may find it easier to live out the rest of their lives without revealing or admitting this aspect of their identity. This could be dangerous insofar as it artificially relieves the wider society of its usual obligations of understanding and support, creating a new kind of ghetto that once again hides the group from the eyes—and protective social policies—of mainstream society.

It is a delicate matter, then, to balance this danger against the competing vision (explored a few specters back) of new media allowing us slowly and safely to explore multiple aspects of our personal and sexual identities. Once again, the most we can do is to be aware, as individuals and as public servants, of this danger, and to make active efforts to take account of even these relatively invisible minorities in lawmaking and social policy.

A less familiar version of the more general worry about "disembodiment" takes the idea quite literally. With so much emphasis on information transmission and digital media, the physical body itself can begin to seem somewhat unnecessary. Respected scientists such as Hans Moravec speak enthusiastically of a future world in which our mental structures are somehow preserved as potentially immortal patterns of information capable of being copied from one electronic storage medium to another. In the reducing heat of such a vision, the human body (in fact, any body, biological or otherwise) quickly begins to seem disposable—"mere jelly" indeed.[38]

To be fair, Moravec himself repeatedly stresses the symbiotic nature of good forms of human-machine relationship. His vision of the self as a kind of persisting higher-order pattern is, ultimately, much more subtle and interesting than his critics allow. But what I seek to engage here is not the true vision but the popular caricature: the idea that the body and its capabilities are fundamentally irrelevant to the mind and hence the self. Nothing, absolutely nothing, in the account I have developed lends support to such a vision of essential disembodiment. In depicting the intelligent agent as a joint function of the biological brain, the rest of the human body *and* the tangled webs of technological support, I roundly reject the vision of the self as a kind of ethereal, information-based construct. There is no informationally constituted *user* relative to whom all the rest is just *tools*. It is, as we argued in chapter 5, *tools all the way down*. We are just the complex, shifting agglomerations of "our own" inner and outer tools for thought. We are our own best artifacts, and always have been.

Some of these tools, to be sure, help constitute our conscious minds, while many operate below or beneath or otherwise outside of that domain. As we have repeatedly seen, it would be crazy to identify the physical basis of oneself solely with the machinery of the conscious goings-on. As we saw, just about everything we do and think arises from a complex interplay between the contents of conscious awareness and reflection and the more subterranean processing that throws up ideas, and supports fluent real-world action. If there is any truth at all, then, in the image of the self as a kind of higher-level pattern, it is a pattern determined by the activities of multiple conscious and nonconscious elements spread across brain, body, and world.

Fine words indeed. But no consolation, one supposes, to our isolated friend, tapping away at the keyboard late at night, fearful of human contact and aroused only by the occasional warbling of "it's not my fault" emanating from the speakers as the machine crashes for the tenth time that day. While this lifestyle may have more good in it than many critics believe, it is (I submit) a vision of the past. The agenda of human-centered technology differs in just about every respect. In particular, such technologies hold out the promise of more mobility, richer interfaces, and richer interactive support. Far from being stuck in an isolated corner, our hero may find herself engulfed in a mobile, varied, and physically demanding social whirl.

First and foremost, human-centered technology aims to free the user from that whole "box on a tabletop" regime: the regime of sitting, looking at a screen, and interfacing with the digital world using the narrow and demanding channels of keyboard and mouse. Wearable computers, augmented reality displays, and richer interface technologies transform this image beyond recognition. Mobile access to the web will soon be as common as mobile access to a phone line. Keyboard interfaces, of all kinds, will be augmented, and sometimes replaced, by the kinds of rich, analogue interface described in chapter 2. Instead of touching tiny and elusive keys to pull up a menu to select a favorite web site, you might just move a finger to touch an icon that only you can see, hanging in the air about three inches above your eyeline. At first, such augmentations may rely on clumsy spectacle-based displays—but in the end, all the new functionality may be engineered into our eyes themselves.

As a simple taste of this kind of freedom, imagine the probable end point of the cell phone revolution. The receiver will be surgically implanted in order to make fairly direct contact with the auditory nerve or perhaps even the ventral cochlear nucleus. Alerted to an incoming call by a characteristic tingling in the fingers, you can take the call without anyone else hearing; your replies need not be spoken aloud as long as you gently simulate the correct muscle movements in your throat and larynx. Such a technology would look, to us today, like to some kind of "telepathy." There are pros and cons to such a scenario, without a doubt, but there would certainly be no feeling of being trapped, bound, or isolated courtesy of such mobile, easy-to-use, communication-extending enhancements.

The point about mobility is probably crucial. Wearable computing and ubiquitous computing are each, in different but complementary ways, geared to freeing the user from the desktop or laptop. They are geared to *matching* the technology to a mobile, socially interactive, physically engaged human life form. The development of new and richer interfaces goes hand in hand with this. The ubiquitous devices will be more self-sufficient—more likely to monitor us than to receive deliberate commands and inputs. We will still need to communicate data and requests at times, and here the use of a variety of physical embodiment-exploiting interfaces will be crucial. The violinist Yo-Yo Ma's communications with his instrument via the bow are, we saw in chapter 2, amazingly rich and nuanced. One day soon we will see expert web-surfers and designers able to manipulate data streams and virtual objects with all the skill and subtlety of a Yo-Yo Ma. Almost certainly, they will not be using a keyboard and mouse to do so.

Where some fear disembodiment and social isolation, I anticipate *multiple* embodiment and social *complexity*. An individual may identify himself as a member of a wide variety of social groups, and may (in part courtesy of the new technologies of telepresence and telerobotics) explore in each of those contexts, a variety of forms of embodiment, contact, and sexuality. The feeling of disembodiment arises only when we are digitally immersed but lack the full spectrum of rich, real-time feedback that body and world provide. As feedback links become richer and more varied, our experience will rather become one of *multiple ways of being embodied*; akin, perhaps, to the way a skilled athlete feels when she exchanges tennis racket for wetsuit and flippers. In these new worlds, Katherine Hayles notes, it is "not a question of leaving the body behind but rather of extending embodied awareness in highly specific local and material ways that would be impossible without electronic prostheses."[39]

In a strange way, we may even come to better appreciate the value and significance of our normal bodily presence by exploring such alternatives. Not disembodiment, then, so much as a deeper understanding of why the body matters and of the space of possible bodies and perspectives. Not isolation so much as a wider and less geocentric kind of community. Not handcuffed to a desktop device in a dusty corner, but walking and running out in the real world. Not mediated via the narrow and distressing bottlenecks of keyboard and screen, but richly coupled via new interfaces that make the most of our biological senses and native bodily skills.

But let's not fool ourselves. The problems all too briefly scouted above are real and pressing, and the solutions I have gestured at are at best partial and often visibly inadequate. Still, there is no turning back. The drive toward biotechnological merger is deep within us—it is the direct expression of much of what is most characteristic of the human species. The task is to merge gracefully, to merge in ways that are virtuous, that bring us closer to one another, make us more tolerant, enhance understanding, celebrate embodiment, and encourage mutual respect. If we are to succeed in this important task, we must first understand ourselves and our complex relations with the technologies that surround us. We must recognize that, in a very deep sense, we were always hybrid beings, joint products of our biological nature and multilayered linguistic, cultural, and technological webs. Only then can we confront, without fear or prejudice, the specific demons in our cyborg closets. Only then can we actively structure the kinds of world, technology, and culture that will build the kinds of *people* we choose to be.

CHAPTER 8

Conclusions: Post-Human, Moi?

The human brain is nature's great mental chameleon. Pumped and primed by native plasticity, it is poised for profound mergers with the surrounding web of symbols, culture, and technology. Human thought and reason emerges from a nest in which biological brains and bodies, acting in concert with nonbiological props and tools, build, benefit from, and then rebuild an endless succession of designer environments. In each such setting our brains and bodies couple to new tools, yielding new extended thinking systems. These new thinking systems create new waves of designer environments, in which yet further kinds of extended thinking systems emerge. By this magic, seeded long ago by the emergence of language itself, the ratchets engage and the golden machinery of mind-design, mind redesign, and mind re-redesign, rumbles into life.

The process continues, and it is picking up speed. Some of our best new tools adapt to individual brains during use, thus speeding up the process of mutual accommodation beyond measure. Human thought is biologically and technologically poised to explore cognitive spaces that would remain forever beyond the reach of non-cyborg animals. Our technologically enhanced minds are barely, if at all, tethered to the ancestral realm. As William Burroughs put it, "We're here to go."[1]

That gravely voice, intoning its insistent verse ("We're here to go. That's what we're here for. We're here to go. . . .") brings us full circle. For Burroughs's sights, like those of the scientists who first coined the term "cyborg," were set on space and on the colonization of other planets. The most significant twenty-first-century frontiers, however, are those not of space but of the mind. Our most significant technologies are those that allow our *thoughts* to go where no animal thoughts have gone before. It is our shape-shifter minds, not our space-roving bodies, that will most fully express our deep cyborg nature.

The word *cyborg* once conjured visions of wires and implants, but as we have seen, the use of such penetrative technologies is inessential. To focus on them is to concede far too much to the ancient biological skin-bag. What matters most is our obsessive, endless weaving of biotechnological webs: the constant two-way traffic between biological wetware and tools, media, props, and technologies. The very best of these resources are not so much used as incorporated into the user herself. They fall into place as aspects of the thinking process. They have the power to transform our sense of self, of location, of embodiment, and of our own mental capacities. They impact who, what and where we are.

In embracing our hybrid natures, we give up the idea of the mind and the self as a kind of wafer-thin inner essence, dramatically distinct from all its physical trappings. In place of this elusive essence, the human person emerges as a shifting matrix of biological and nonbiological parts. The self, the mind, and the person are no more to be extracted from that complex matrix than the smile from the Cheshire Cat.

Some fear, in all this, a loathsome "post-human" future.[2] They predict a kind of technologically incubated mind-rot, leading to loss of identity, loss of control, overload, dependence, invasion of privacy, isolation, and the ultimate rejection of the body. And we *do* need to be cautious, for to recognize the deeply transformative nature of our biotechnological unions is at once to see that not all such unions will be for the better. But if I am right—if it is our basic *human* nature to annex, exploit, and incorporate nonbiological stuff deep into our mental profiles—then the question is not whether we go that route, but in what ways we actively sculpt and shape it. By seeing ourselves as we truly are, we increase the chances that our future biotechnological unions will be good ones.

Notes

Introduction

1. I first encountered this example in D. Rumelhart, P. Smolensky, D. McClelland, and G. Hinton, "Schemata and Sequential Thought Processes in PDP Models," vol. 2 of *Parallel Distributed Processing: Explorations in the Microstructure of Cognition* (Cambridge, Mass.: MIT Press, 1986), 7–57.

2. For a brief sampling, see the essays in *The Adapted Mind*, ed. J. Barklow, L. Cosmides, and J. Tooby (New York: Oxford University Press, 1992).

3. See W. J. Holstein "Moving Beyond the PC," *US News and World Report* 127:23 (December 13, 1999): 49–58. Jacob Mey (personal communication) tells me that the word "kanny" is an abbreviated version of a word used, when talking to small children, to mean (in a sweet way) "palm of the hand." This is combined with an ending that signifies nounhood and conveys the status of an artifact. The best translation he can suggest is "palmie."

Chapter 1

1. M. Clynes and N. Kline, "Cyborgs and Space," *Astronautics*, September 1960; reprint, *The Cyborg Handbook*, ed. by C. Gray (London: Routledge, 1995), 29–34.

2. Ibid., 29.

3. See A. Turing, "On Computable Numbers, with an Application to the Entscheidungsproblem," *Proceedings of the London Mathematical Society*, 2d ser., 42 (1936): 230–65; A. Turing, "Computing Machinery and Intelligence," *Mind*

59 (1950): 433–60; J. Von Neumann, *The Computer and the Brain* (New Haven: Yale University Press, 1958); W. S. McCulloch and W. H. Pitts, "A Logical Calculus of the Idea Immanent in Nervous Activity," *Bulletin of Mathematical Biophysics* 5 (1943): 115–33; A. Newell, J. Shaw, and H. Simon, "Empirical Explorations with the Logic Theory Machine," *Proceedings of the Western Joint Computer Conference* 15 (1957): 218–39.

4. See R. Ashby, *Introduction to Cybernetics* (New York: Wiley, 1956), and N. Wiener, *Cybernetics, or Control and Communication in the Animal and in the Machine* (New York: Wiley, 1948).

5. Manfred Clynes, "An Interview with Manfred Clynes," interview by C. H. Gray, *The Cyborg Handbook* (London: Routledge, 1995), 43–55.

6. Clynes and Kline, "Cyborgs and Space."

7. Donna Haraway, "Cyborgs and Symbionts," *The Cyborg Handbook,* ed. C. Gray (London: Routledge, 1995), xv.

8. Simon Le Vay, "Brain Invaders," *Scientific American* 282:3 (March 2000): 27.

9. K. Warwick, "Cyborg 1.0," *Wired* 8:2 (February 2000): 145.

10. The experiments and results are further detailed in Professor Kevin Warwick's book *I, Cyborg,* (London: Century Press, 2002).

11. Warwick, "Cyborg 1.0," 146–47.

12. All these cases are reported by Kevin Warwick, *Wired,* 150. For fuller discussion, see chapter 5.

13. Ian Sample, "Push My Button," reporting on work by Stuart Meloy, a surgeon at Piedmont Anesthesia and Pain Consultants in North Carolina. See *New Scientist,* February 10, 2001, 23.

14. See C. H. Gray ed., *The Cyborg Handbook,* "Pilot's Associate," 101–3, and "Science Fiction Becomes Military Fact," 104–5.

15. For "adamantium skeleton," see the Marvel Comics Character: Wolverine. For "skull guns" and "human brains directly jacked into Cyberspace," see W. Gibson, *Neuromancer* (New York: Ace Books, 1984).

16. W. Ellis, *Transmetropolitan,* 3, Helix DC Comics (New York, 1997), 4. (Bar-coded dancer appears on page 4.)

17. For some nice discussions, see Kim Sterelny and Paul Griffiths, *Sex and Death* (Chicago: University of Chicago Press, 1999).

18. As Kevin Warwick once usefully remarked (personal conversation).

19. E. Hutchins, "How a Cockpit Remembers Its Speeds," *Cognitive Science* 19 (1995): 265–88; E. Hutchins and T. Klansen, "Distributed Cognition in an Airline Cockpit," in *Cognition and Communication at Work,* ed. Y. Engestrom and D. Middleton (Cambridge: Cambridge University Press, 1998); E. Hutchins, "Inte-

grated Mode Management Interface," *Final Report to Contract NCC-2-591 NASA-Ames Research Center* (Moffert Field, Calif., 1997).

20. L. Zuckerman, "Making Computers Relate to Their Human Partners," *New York Times*, March 4, 2000. Available on the web at http://www.nytimes.com/00/03/04/news/arts/machines-humans.html.

21. Ibid.

22. Kevin Kelly, *Out of Control* (Reading, Mass.: Perseus Books, 1994), 331.

23. Clynes and Kline, "Cyborgs and Space."

24. Nicola Jones, "Call from the Heart," *New Scientist* 2277 (February 10, 2001): 20.

25. In Finland, a service is being pioneered that allows you to use your cell phone to monitor the actual (not the timetabled) approach of your bus to a designated bus stop, setting it up to alert you when it is time to leave your warm surroundings and venture into the cold Finnish night (service pioneered by Hewlett Packard—see www.hp.com). For a variety of reasons, then, the cell phone is unusually well poised to act as a transition technology. In the wake of the September 11 tragedy, many Americans now view such items as necessities, highlighting its role in informing rescuers, exchanging vital last-minute information, and (sadly) conveying a final message of love from those about to die. See Olivia Baker "Cellphones Hit Home," *USA Today*, September 13, 2001, 12D.

26. Geoff Marsh, "Make a Connection Without the Clutter," *Sunday Express*, October 15, 2000.

27. Mark Weiser, "The Computer for the 21st Century," *Scientific American*, September 1991, 94–110.

28. See Don Norman, *The Invisible Computer* (Cambridge, Mass.: MIT Press, 1999).

29. See S. Quartz and T. Sejnowski, "The Neural Basis of Cognitive Development: A Constructivist Manifesto," *Behavioral and Brain Sciences* 20 (1997), 537–96.

30. For a recent account, see D. Milner and M. Goodale, *The Visual Brain in Action* (Oxford: Oxford University Press, 1995). This work is further discussed in chapters 4 and 7, and in A. Clark "Visual Awareness and Visuomotor Action," *Journal of Consciousness Studies* 6:11–12 (1999): 1–18.

31. Some names to conjure with (past and present) include Lev Vygotsky, Maurice Merleau-Ponty, Jerome Bruner, Bruno Latour, Daniel Dennett, Ed Hutchins, Don Norman, and (to a greater or lesser extent) all those currently working in the field of "situated and distributed cognition."

Chapter 2

1. You can find a nice description on the web at www.wps.com/about-WPS/ personal/black-hole/.

2. See Mark Weiser, "The Computer for the 21st Century," *Scientific American*, September 1991, 94–110, and Donald Norman, *The Invisible Computer* (Cambridge, Mass.: MIT Press, 1999). Similar ideas are found in the philosophical works of Maurice Merleau-Ponty, such as *The Phenomenology of Perception*, trans. Paul Kegan (1945; reprint, London: Routledge, 1962) and Martin Heidegger, *Being and Time* (1927; reprint, New York: Harper and Row, 1961). An early appearance of the idea of "transparent tools" in the context of human-computer interaction is found in Jacob Mey's 1988 paper "CAIN and the Transparent Tool: Cognitive Science and the Human-Computer Interface" presented at the 3rd Symposium on Human Interface in Osaka, Japan, in *Journal of the Society of Instrument and Control Engineers* 27:1 (1987): 247–52.

3. See Donald Norman, *The Invisible Computer* (Cambridge, Mass.: MIT Press, 1999).

4. There is some discussion of this case in Norman, but what follows is based upon David Landes's wonderful tome *Revolution in Time: Clocks and the Making of the Modern World* (London: Viking Press, 2000).

5. Ibid., 92–93.

6. I came upon this example several years ago but cannot seem to recover the original source. To my best recollection, the example was then deployed in the service of a philosophical argument concerning the nature of implicit beliefs. If any reader can supply the full reference, I would be most grateful.

7. A tropism is a kind of automatic, hardwired response. I use the term, however, to mean a learned but now largely automatic response.

8. For a sustained philosophical defense of this position, see A. Clark and D. Chalmers, "The Extended Mind," *Analysis* 58 (1998): 7–19.

9. This idea of "scaffolding" originates with the work of Soviet psychologist Lev Vygotsky, who stressed the way a child's experience with external props (especially an adult's helps and hints) could alter and inform the way the child solves a problem. Since then, the term "scaffolding" has come to mean any kind of external aid and support, whether provided by a notepad, a computer, or another human being. See L. Vygotsky, *Thought and Language* (Cambridge, Mass.: MIT Press translations, 1986).

10. Norman credits the term to Jeff Raskin and dates the coinage as 1978, in an internal Apple memo. See Norman, *The Invisible Computer*, 275, par.1.

11. This list is a distillation from Norman's work. It is also influenced by conversations with Ed Hutchins at UCSD and Mike Scaife and Yvonne Rogers at the University of Sussex.

12. Norman, *The Invisible Computer*, 59.

13. Ibid., 67.

14. For this, see Bradley J. Rhodes, Nelson Minar, and Josh Weaver, "Wearable Computing Meets Ubiquitous Computing: Reaping the Best of Both Worlds" (*Proceedings of the International Symposium on Wearable Computers*, October 1999), http://www.media.mit.edu/rhodes/Papers/wearhive.html.

15. Bradley J. Rhodes, "The Wearable Remembrance Agent: A System for Augmented Memory," *Personal Technologies Journal Special Issue on Wearable Computing, Personal Technologies* 1 (1997): 218.

16. Ibid.

17. Known as a "Private Eye," the unit is made by Reflection Technology and marketed (at last sighting) by Phoenix Group, N.Y. See www.reflection.com for details.

18. See T. Starner, S. Mann, B. Rhodes, J. Levine, J. Healey, D. Kirsch, R. W. Picard, and A. P. Pentland, "Augmented Reality through Wearable Computing," *Presence, Special Issue on Augmented Reality* (1997).

19. P. Scott, "Eye Spy," *Scientific American*, September 2001, News Scan sec.

20. Rhodes, "The Wearable Remembrance Agent," 219.

21. For much more on this issue, see chapter 7.

22. Paul Dourish, *Where the Action Is: The Foundations of Embodied Interaction* (Cambridge, Mass.: MIT Press, 2001), 141.

23. Ibid., 42–43.

24. Ibid., 42.

25. Thanks to Ron Chrisley for pointing this out.

26. J. Patten, H. Ishii, J. Hines, and G. Pangaro, "Sensetable: A Wireless Object Tracking Platform for Tangible User Interfaces" (Proceedings of the ACM CHI 2001), http://citeseer.nj.nec.com/464521.html.

27. B. Ullmer and H. Ishii, "The metaDESK: Models and Prototypes for Tangible User Interfaces" (Proceedings of the ACM UIST '97 Symposium on User Interface Software and Technology, 1997), 223–32. http://citeseer.nj.nec.com/ullmer97metadesk.html.

28. See especially David Small and Hiroshi Ishii, "Design of Spatially Aware Graspable Displays" (Short Paper, during proceedings of the ACM CHI '97, Atlanta, March 22–27, 1997).

29. Thanks to Frank Biocca for helpful discussion on this topic.

30. For this information and the examples in the same paragraph, I am indebted to S. Feiner, "Augmented Reality: A New Way of Seeing," *Scientific American* 286:4 (April 2002): 48–55.

31. The idea of using mixed reality play was introduced to me by Steve Benford and Tom Rodden in a plenary session at the Fourth International Conference of Cognitive Technology, Coventry, UK, 2001. There, they described a "traveling story tent," using some of the technologies described in the text.

32. Yvonne Rogers, Mike Scaife, Eric Harris, Ted Phelps, Sara Price, Hilary Smith, Henk Muller, Cliff Randall, Andrew Moss, Ian Taylor, Danae Stanton, Claire O'Malley, Greta Corke, and Silvia Gabriella, "Things Aren't What They Seem to Be: Innovation Through Technology Inspiration" (unpublished manuscript).

33. One highly funded European endeavor, the Equator Project, is entirely dedicated to the exploration of such interpenetration. The Equator Project presently spans eight academic institutions. A typical vision statement reads: "Instead of treating these [the digital and the physical] as two different worlds, our view is of these two media as being complementary halves of the same world. They are continually interwoven by the activity of people." For more on this work, visit the Equator web site: www.equator.ac.uk.

Chapter 3

1. V. S. Ramachandran and S. Blakeslee, *Phantoms in the Brain: Probing the Mysteries of the Human Mind* (New York: William Morrow, 1998), 58.

2. Ibid., 59

3. Galvanic Skin Response (GSR) is basically a measure of arousal. Arousal (either positive or negative) is accompanied by increased blood flow, heart rate, and sweating. Experimenters measure changes in the electrical resistance of the skin, caused by the sweating, to yield an index of arousal.

4. Ramachandran and Blakeslee, *Phantoms in the Brain*, 52–54.

5. Ibid., 40–42. A genetic component is suggested by, for example, the presence of phantom arms and hands in a patient born without arms.

6. Ibid., 62.

7. Y. Iwamura, "Hierarchical Somatosensory Processing," *Current Opinion in Neurobiology* 8 (1998): 522–28.

8. A. Yarbus, *Eye Movements and Vision* (New York: Plenum Press, 1967).

9. http://members.tripod.com/andybauch/magic.html, or just feed "amazing card trick" to a search engine such as Google.

10. See also Daniel Dennett's "many Marilyns" example, as described in D. Dennett, *Consciousness Explained* (Boston: Little, Brown, 1991).

11. G. W. McConkie and K. Rayner, "Asymmetry of the Perceptual Span in Reading," *Bulletin of the Psychonomic Society* 8 (1976): 365–68.

12. J. K. O'Regan, "Solving the 'Real' Mysteries of Visual Perception: The World as an Outside Memory," *Canadian Journal of Psychology* 46 (1992): 461–88.

13. D. J. Simons and D. T. Levin, "Change Blindness," *Trends in Cognitive Science* 1 (1997): 261–67.

14. Ibid., 266.

15. Best just to search for Change Blindness, Flicker Paradigm, etc., but some current sites are

http://nivea.psycho.univ-paris5.fr

http://coglab.wjh.harvard.edu

http://www.cbr.com/rensink

16. R. Brooks, "Intelligence Without Representation," *Artificial Intelligence* 47 (1991):139–59.

17. A. Clark, *Being There: Putting Brain, Body and World Together Again* (Cambridge, Mass.: MIT Press, 1997); D. Dennett, *Consciousness Explained* (Boston: Little, Brown, 1991); D. Ballard, "Animate Vision," *Artificial Intelligence* 48 (1991): 57–86; P. S. Churchland, V. S. Ramachandran, and T. Sejnowski, "A Critique of Pure Vision," in *Large-Scale Neuronal Theories of the Brain*, ed. C. Koch and J. Davis (Cambridge, Mass.: MIT Press, 1994).

18. With the important exception of the work of Alva Noe and J. Kevin O'Regan, "A Sensorimotor Account of Vision and Visual Consciousness," *Behavioral and Brain Sciences* 24 (2001): 5.

19. For a fairly conservative view see S. Pinker, *Words and Rules* (New York: Basic Books, 1999). For radical views more in line with the one I develop here, see D. Dennett, *Consciousness Explained* (Boston: Little, Brown, 1991); R. Jackendoff, "How Language Helps Us Think," *Pragmatics and Cognition* 4:1 (1996): 1–34; L. S. Vygotsky, *Thought and Language* (Cambridge, Mass.: MIT Press, 1986). I explore these topics more fully in A. Clark, "Magic Words: How Language Augments Human Computation," in *Thought and Language*, ed. S. Boucher and P. Carruthers (Cambridge: Cambridge University Press, 1998), and A. Clark, *Being There*.

20. In a recent treatment Merlin Donald suggests that prior to the ability to acquire complex language and culture comes an ability to learn complex sequences of connected subskills (something like the automated subroutines of a computer program). This ability, he argues, is rooted in a change in the nature and functional role of consciousness itself. See Merlin Donald, *A Mind So Rare* (London: Routledge, 2001).

21. R. K. R. Thompson, D. L. Oden, and S. T. Boysen, "Language-Naive Chimpanzees (*Pan troglodytes*) Judge Relations Between Relations in a Conceptual Matching-to-sample Task," *Journal of Experimental Psychology: Animal Behavior Processses* 23 (1997): 31–43.

22. The power of labeling is celebrated in D. Dennett, "Learning and Labeling: Commentary on A. Clark and A. Karmiloff-Smith," *Mind & Language* 8 (1994): 540–548.

23. Oliver Sacks, *Seeing Voices* (New York: Harper Perennial, 1989).

24. S. Dehaene, E. Spelke, P. Pinel, R. Stanescu, and S. Tviskin, "Sources of Mathematical Thinking: Behavioral and Brain Imaging Evidence," *Science* 284 (1999): 970–74. See also Dehaene's superb book, *The Number Sense* (Oxford: Oxford University Press, 1997).

25. Dehaene, *The Number Sense*, 103.

26. Some newer kinds of computational models, so-called connectionist systems or Artificial Neural Networks, display a similar profile of strengths and weaknesses, making them a much better bet as machine models of some of the biological aspects of intelligence. These models have been the focus of much of my own past research. See A. Clark, *Microcognition: Philosophy, Cognitive Science and Parallel Distributed Processing* (Cambridge, Mass.: MIT Press, 1989). For good original sources, see *Parallel Distributed Processing: Explorations in the Microstructure of Cognition*, ed. J. McClelland, D. Rumelhart, and P. R. Group (Cambridge, Mass.: MIT Press/Bradford Books, 1986), vols. I & II. Most older A.I. (Artificial Intelligence) models, by contrast, seem much more like models of the larger systems (of humans-plus-technological props and aids) on which they were originally based. For a nice account, see E. Hutchins, *Cognition in the Wild* (Cambridge, Mass.: MIT Press, 1995).

27. C. Van Leeuwen, I. Verstijnen, and P. Hekkert. (1999). "Common Unconscious Dynamics Underlie Common Conscious Effects: A Case Study in the Interactive Nature of Perception and Creation." In *Modelling Consciousness Across the Disciplines,* ed. J.S. Jordan (Lanhan, Md.: University Press of America), 179–218.

28. D. Chambers and D. Reisberg. "Can Mental Images Be Ambiguous?" *Journal of Experimental Psychology: Human Perception and Performance* 2:3 (1985): 317–28.

29. Van Leeuwen et al. (See note 27.)

30. Hutchins, *Cognition in the Wild*, 155.

31. For more on all this, see Clark, "Magic Words" (refer to note 19); A. Clark and C. Thornton, "Trading Spaces: Connectionism and the Limits of Uninformed

Learning," *Behavioral and Brain Sciences* 20:1 (1997): 57–67; D. Dennett, *Kinds of Minds* (New York: Basic Books, 1996); M. Donald, *Origins of the Modern Mind* (Cambridge, Mass.: Harvard University Press, 1991).

32. Dennett, *Kinds of Minds*, 133.

33. Donald, *Origins of the Modern Mind*, 343.

34. See V. Landi, *The Great American Countryside* (New York: Collier Macmillan, 1982), 361–63.

35. J. Elman, "Learning and Development in Neural Networks: The Importance of Starting Small," *Cognition* 48 (1994): 71–99.

36. S. Fahlman and C. Lebiere, "The Cascade-Correlation Learning Architecture," in *Advances in Neural Information Processing Systems 2*, ed. D. Touretzky (San Francisco: Morgan Kauffman, 1990); C. Thornton, *Truth from Trash* (Cambridge, Mass.: MIT Press, 2000).

37. Specifically, synaptic and dendritic growth. See S. Quartz and T. Sejnowski, "The Neural Basis of Cognitive Development: A Constructivist Manifesto," *Behavioral and Brain Sciences* 20 (1997): 537–96.

38. Evidence for the view comes primarily from recent neuroscientific studies (especially work in developmental cognitive neuroscience). Key studies here include work involving cortical transplants, in which chunks of visual cortex were grafted onto other cortical locations (such as somatosensory or auditory cortex) and proved plastic enough to develop the response characteristics appropriate to the new location. See B. Schlagger and D. O'Leary, "Potential of Visual Cortex to Develop an Array of Functional Units Unique to Somatosensory Cortex," *Science* 252 (1991): 1556–60, as well as work showing the deep dependence of specific cortical response characteristics on developmental interactions between parts of the cortex and specific kinds of input signals; see A. Chenn et al., "Development of the Cerebral Cortex," in *Molecular and Cellular Approaches to Neural Development*, ed. W. Cowan, T. Jessel, and S. Ziputsky (Oxford: Oxford University Press, 1997), 440–73). There is also, as mentioned earlier, a growing body of constructivist work in Artificial Neural Networks, connectionist networks in which the architecture (number of units and layers, etc.) itself alters as learning progresses. The take-home message is that immature cortex is surprisingly homogeneous, and that it "requires afferent input, both intrinsically generated and environmentally determined, for its regional specialization." See S. Quartz, "The Constructivist Brain," *Trends in Cognitive Sciences* 3:2 (1999): 48–57.

39. R. Sireteanu, "Switching On the Infant Brain," *Science* 286 (1999): 60.

40. P. Griffiths and K. Stotz, "How the Mind Grows: A Developmental Perspective on the Biology of Cognition," *Synthese* 122 (2000): 1–2, 29–52.

41. This notion was partially anticipated by Gregory Bateson's notion of an "extra-regulator": a creature that regulates its own states by changing and controlling its environment. Bateson's concern, like that of Clynes and Kline (see chapter 1), was more with basic bodily functions than with mental processes. See Bateson, "The Role of Somatic Change in Evolution" (1972); reprint, in his *Steps to an Ecology of Mind* (Chicago: University of Chicago Press, 2000), 346–63.

42. For a powerful defense of such a view, see P. Griffiths and R. Gray, "Developmental Systems and Evolutionary Explanation," *Journal of Philosophy* 91: 6 (1994): 277–305.

Chapter 4

1. The friend was Brian Cantwell Smith, now a professor at Duke University, and former director of the XeroxPARC component of the Center for the Study of Language and Information at Palo Alto, California.

2. D. Dennett, "Where Am I?" in D. Dennett, *Brainstorms* (Sussex: Harvester Press, 1981).

3. Ibid., 317.

4. See Johan Wessberg, Christopher R. Stambaugh, Jerald D. Kralic, Pamela D. Beck, Mark Laubach, John K Chapin, Jung Kim, S. Jammes Biggs, Mandayam A. Srinivasan, and Miguel A. L. Nicolelis, "Real-time Prediction of Hand Trajectory by Ensembles of Cortical Neurons in Primates," *Nature* 408 (November 16, 2000): 305–6.

5. The official name is the Laboratory for Human and Machine Haptics.

6. Quoted in MIT Tech Talk (Cambridge, Mass.: MIT news office, December 6, 2000); http://web.mit.edu/newsoffice.

7. This point is nicely made in David Sanford's follow-up to Dennett's piece, "Where Was I," in *The Mind's I*, ed. D. Dennett and D. Hofstadter (Sussex: Harvester Press, 1981), 232–41.

8. In an article published in *Omni* (May 1980): 45–52, Minsky credits Pat Gunkel with the original coinage.

9. I was led to this site by Thomas Campanella's excellent piece "Eden by Wire," in *The Robot in the Garden*, ed. K. Goldberg (Cambridge, Mass.: MIT Press, 2000).

10. Marvin Minsky, *Omni* (May 1980): 45–52. The passage is cited in Dennett and Hofstadter, *The Mind's I*.

11. A. Hein, "The Development of Visually-Guided Behavior," in *Visual Coding and Adaptability*, ed. C. S. Harris (Hillsdale, N.J.: Erlbaum, 1980), 52. For further

discussion and some caveats, see S. Hurley, *Consciousness in Action* (Cambridge, Mass.: Harvard University Press, 1998), chaps. 9, 10.

12. See J. G. Taylor, *The Behavioral Basis of Perception* (New Haven, Conn.: Yale University Press, 1962), 205, and Hurley, *Consciousness in Action*, 387.

13. These experiments were performed by some of my ex-colleagues at the Washington University Medical School. See W. Thach, H. Goodkin, and J. Keating, "The Cerebellum and the Adaptive Coordination of Movement," *Annual Review of Neuroscience* 15 (1992): 403–42.

14. V. S. Ramachandran and S. Blakeslee, *Phantoms in the Brain: Probing the Mysteries of the Human Mind* (New York: William Morrow, 1998), 59.

15. This list is based on K. Goldberg, ed., "Introduction: The Unique Phenomenon of a Distance," *The Robot in the Garden* (Cambridge, Mass.: MIT Press, 2000).

16. Ibid. See all the papers therein, especially Machiko Kusahara's survey, "Presence, Absence and Knowledge in Telerobotic Art."

17. E. Kac, "Dialogical Telepresence and Net Ecology," in Goldberg, *The Robot in the Garden*, 188.

18. The next few paragraphs draw mainly on Blake Hannaford's "Feeling Is Believing: A History of Telerobotics," in Goldberg, *The Robot in the Garden*, 247–74.

19. R. C. Goertz, "Fundamentals of General-Purpose Remote Manipulators," *Nucleonics* 10:11 (November 1982): 36–45.

20. A. Bejczy and K. Salisbury "Kinesthetic Coupling for Remote Manipulators," *Computers in Mechanical Engineering* 2:1 (1983): 48–62.

21. Hannaford "Feeling Is Believing," in Goldberg, *The Robot in the Garden*, 251.

22. This work is described in A. Milner and M. Goodale, *The Visual Brain in Action* (Oxford: Oxford University Press, 1995), 167–70, and in M. Gazzaniga, *The Mind's Past* (Berkeley: University of California Press, 1998), 106–10.

23. A large literature, pro and con, has grown around this dramatic demonstration. E. Brenner and J. Smeets, "Size Illusions Influence How We Read but Not How We Grasp an Object," *Experimental Brain Research* 111 (1996): 473–76; R. Ellis, J. Flanagan, and S. Lederman, "The Influence of Visual Illusions on Grasp Position," *Experimental Brain Research* 125 (1999): 109–14; Gazzaniga, *The Mind's Past*, chap. 5; Ramachandran and Blakeslee, *Phantoms in the Brain*, chap. 4; S. Glover, "Visual Illusions Affect Planning but Not Control," *Trends in Cognitive Sciences* 6:7 (2002): 288–92.

24. In fact, the two visual "brains" are sufficiently distinct as to be independently vulnerable to damage. DF, a patient who suffered severe damage (due to carbon monoxide poisoning) to the ventral stream, claims she is unable to see the shape or orientation of visually presented objects. Yet, if you ask her to drop a

letter through a mail slot (which she says she cannot see), she will do so accurately. Optic ataxics, by contrast, have damage to the dorsal stream and are unable to perform fluent motor actions despite seeing the scene perfectly well and suffering no gross physical impediments to fluent action. See Milner and Goodale, *The Visual Brain*.

25. T. Sheridan, *Telerobotics, Automation and Human Supervisory Control* (Cambridge, Mass.: MIT Press, 1992).

26. M. Goodale, "Where Does Vision End and Action Begin?" *Current Biology* R489–R491 (1998): 491.

27. Gazzaniga, *The Mind's Past*, 106.

28. Hannaford, "Feeling Is Believing," in Goldberg, *The Robot in the Garden*, 255.

29. Perhaps there is a *sense* in which it *was*, but the signals ran *through* the helper whose concealed tappings completed the circuit.

30. Ramachandran and Blakeslee, *Phantoms in the Brain*, 61.

31. Ibid.

32. F. Biocca, and J. Rolland, "Virtual Eyes Can Rearrange Your Body: Adaptation to Visual Displacement in See-through, Head-mounted Displays," *Presence: Teleoperators and Virtual Environments* (Cambridge, Mass.: MIT Press, 1998), 7: 3, 262–77.

33. See Antonio Damasio, *Descartes' Error* (New York: Grosset Putnam, 1994), 62–66.

34. See M. Kawato et al., "A Hierarchical Neural Network Model for the Control and Learning of Voluntary Movement," *Biological Cybernetics* 57 (1987): 169–85; P. Dean, J. Mayhew, and P. Langdon, "Learning and Maintaining Saccadic Accuracy," *Journal of Cognitive Neuroscience* 6 (1994): 117–38. I first learned about this work from Rick Grush. See R. Grush, "The Architecture of Representation," *Philosophical Psychology* 10:1 (1997): 5–25.

35. See Thach et al., "The Cerebellum and the Adaptive Coordination of Movement," *Annual Review of Neuroscience* 15 (1992): 403–42.

36. W. Kim and A. Bejczy, "Demonstration of a High-Fidelity Predictive/Preview Display Technique for Telerobotic Servicing in Space," *IEEE Trans. Robotics and Automation* 9:5 (1993): 698–702.

37. For an excellent survey, upon which much of the previous discussion is based, see Hannaford, "Feeling Is Believing," in Goldberg, *The Robot in the Garden*.

38. Ibid., 274.

39. Jim Hollan and Scott Stormetta, *Proceedings of the ACM (Association For Computing Machinery)*, ACM 0-89791-S513-S/92/0005-0119 (1992): 119–25.

40. Ibid., 120.

41. Ibid., 125

42. This term was coined by Howard Rheingold in his classic *Virtual Reality* (London: Seiter and Warburg, 1991.)

43. Hubert Dreyfus, "Telepistemology: Descartes's Last Stand," in Goldberg, *The Robot in the Garden*, offers a balanced and sophisticated treatment of such worries.

44. If your partner was very familiar to you, an emulation circuit might help here, but that seems a little dramatic, even for my liberal tastes.

45. Albert Borgman, "Information, Nearness and Farness," in Goldberg, *The Robot in the Garden*, calls this property of endless richness "repleteness."

46. Dreyfus "Telepistemology," in Goldberg, *The Robot in the Garden*, 62.

47. A. Chang, B. Resner, B. Koerner, X. Wang, and H. Ishii, "LumiTouch: An Emotional Communication Device" (short paper), in Extended Abstracts of Conference on Human Factors in Computing Systems (CHI '01), (Seattle, Washington, USA, March 31–April 5, 2001), (New York: ACM Press), 313–14.

48. See S. Brave, A. Dahley, P. Frei, V. Su, and H. Ishii, "inTouch" in *SIGGRAPH* [Special Interest on Computer Graphics] *1998: Conference Abstracts and Applications of Enhanced Realities* (New York: ACM Press, 1998).

49. J. Canny and E. Paulos "Tele-Embodiment and Shattered Presence: Reconstructing the Body for Online Interaction," in Goldberg, *The Robot in the Garden*, 280–81.

50. This theme is also explored by M. Indinopulos "Telepistemology, Mediation and the Design of Transparent Interfaces," in Goldberg, *The Robot in the Garden*.

51. Canny and Paulos, in Goldberg, *The Robot in the Garden*.

52. N. K. Hayles, *How We Became Post-Human* (Chicago: University of Chicago Press, 1999), 291.

Chapter 5

1. EMG stands for Surface Electromyography and involves the use of electrodes, positioned on the body surface, to record information about muscular and nervous activity.

2. From the Stelarc web site at www.stelarc.va.com.au

3. Ibid.

4. Thanks to Blay Whitby, who saw the combined performance, for suggesting this example.

5. See D. Norman, *The Invisible Computer* (Cambridge, Mass.: MIT Press, 1999),7.

6. See D. Norman, *Things That Make Us Smart* (Cambridge, Mass.: Perseus Books, 1993), 191.

7. Ibid., 190.

8. See Johan Wessberg, Christopher R. Stambaugh, Jerald D. Kralic, Pamela D. Beck, Mark Laubach, John K. Chapin, Jung Kim, S. James Biggs, Mandayam A. Srinivasan, and Miguel A. L. Nicolelis, "Real-time Prediction of Hand Trajectory by Ensembles of Cortical Neurons in Primates," *Nature* 408 (November 16, 2000): 305–6.

9. Also called the Laboratory for Human and Machine Haptics, directed by Mandayam Srinivasan.

10. See Bernard D. Reger, Karen M. Fleming, Vittorio Sanguineti, Simon Alford, and Ferdinando A. Mussa-Ivaldi, "Connecting Brains to Robots: The Development of a Hybrid System for the Study of Learning in Neural Tissues," *Artificial Life* 6 (2000): 307–24.

11. T. B. DeMarse, D. A. Wagenaar, A. W. Blau, and S. M. Potter, "The Neurally Controlled Animal: Biological Brains Acting with Simulated Bodies." *Autonomous Robots* 11 (2001): 305–10.

12. N. Birbaumer, N. Ghanayim, T. Hinterberger, I. Iversen, B. Kotchoubey, A. Kübler, J. Perelmouter, E. Taub, and H. Flor, "A Spelling Device for the Paralysed," *Nature* 398 (1999): 297–98.

13. Duncan Graham-Rowe, "Think and It's Done," *New Scientist*, October 17, 1998.

14. In fact, our best current developmental stories suggest that this is precisely the task that confronts the infant. See E. Thelen and L. Smith, *A Dynamic Systems Approach to the Development of Cognition and Action* (Cambridge, Mass.: MIT Press, 1994).

15. William H. Dobelle, "Artificial Vision for the Blind by Connecting a Television Camera to the Visual Cortex," *ASAIO* [American Society of Artificial Internal Organs] *Journal* 46 (2000): 3–9.

16. Commercial concerns, such as Cyberonics Inc., Medtronic Corps., AllHear, and Optobionics Corporation, are already helping to make it happen. This short list was drawn from a helpful article titled "Real-World Cyborgs," which appeared dateline 06/13/00 at www.ai.about.com/compute/ai/weekly/aa061300a.html.

17. Paul Bach-y-Rita, *Brain Mechanisms in Sensory Substitution* (New York: Academic Press, 1972).

18. Paul Bach-y-Rita, Kurt A. Kaczmarek, Mitchell E. Tyler, and Jorge Garcia-Lara, "Form Perception with a 49-point Electrotactile Stimulus Array on the Tongue: A Technical Note," *Journal of Rehabilitation Research and Development* 35:4 (October 1998): 427–30.

19. For more information, see the Epistemics Ltd. web site currently at http://www.epistemics.co.uk/services/projects/foas/.

20. For a nice introductory account, see Michael Gazzaniga, *The Mind's Past* (Berkeley: University of California Press, 1998).

21. D. Norman, *Things That Make Us Smart* (Reading, Mass.: Addison-Wesley, 1993).

22. Ibid., 212–14.

23. The diary entries are based on ideas gathered from many sources, most of which we have already met. The images of information appliances and wearable computers are due to Dan Weiser and Donald Norman. Donald Norman also imagined the neurophone, Stelarc imagined the linked-dancers application, and Patti Maes wrote about brain-like software agents. The thought-controlled prosthesis is based on the work of Nicolelis, Mussa-Ivaldi, Birbaumer, and others described earlier. The "prosthetic car" is inspired by DERA's "Cognitive cockpit."

24. Named after that rather wonderful treatment by Ed Regis, *Great Mambo Chicken and the Transhuman Condition: Science Slightly Over the Edge* (New York: Addison-Wesley, 1990).

25. D. Dennett, *Elbow Room* (Oxford: Oxford University Press, 1984), 82.

26. J. Glover, *I: The Philosophy and Psychology of Personal Identity* (London: Penguin, 1988), 74.

27. On the "narrative self," see D. Dennett, *Consciousness Explained* (Boston: Little, Brown, 1991), chap. 13; A. Damasio, *The Feeling of What Happens* (New York: Harcourt, Brace, 1999), chap. 7.

28. This is not to suppose that some neural circuits have some magic property that makes their goings-on consciously available and others not. It is simply to notice that we are not always conscious of all the processing we are doing, and that some of it is never consciously visible at all. The story I shall develop is thus compatible with both a Dennett-style rejection of any "magic dust" story about consciousness, and with the possibility that conscious experience is nothing but a certain kind of interplay or relationship between multiple nonconscious elements, or a certain kind of informational poise.

29. See A. Clark, "Leadership and Influence: The Manager as Coach, Nanny and Artificial DNA," in *The Biology of Business*, ed. J. Clippinger III (San Francisco: Jossey-Bass, 1999). Also Philip Anderson, "Seven Levers for Guiding the Evolving Enterprise," in Clippinger, *The Biology of Business*.

30. Kevin Kelly, *Out of Control* (Reading, Mass.: Perseus Books, 1994).

31. For this argument, see K. Butler, *Internal Affairs: A Critique of Externalism in the Philosophy of Mind* (Dordrecht: Kluwer, 1998). For some related considerations, see F. Adams and K. Aizawa, "The Bounds of Cognition," *Philosophical Psychology* 14:1 (2001): 43–64.

32. V. S. Ramachandran and S. Blakeslee, *Phantoms in the Brain: Probing the Mysteries of the Human Mind* (New York: William Morrow, 1998). The authors discuss these issues, isolating the anterior cingulate gyrus as one key neural structure.

33. Ibn Sina Avicenna, *De Anima*, vol. 7. Avicenna was a Persian philosopher, scientist, and physician, who lived from 980 to 1037 A.D. See Avicenna Latinus, *Liber de anima seu sextus de naturalibus*. Edition critique de la traduction latine médiévale. Introduction sur la doctrine psychologique d'Avicenne par G. Verbeke, Partes IV–V. The quote is from an unpublished translation by R. Martin.

34. See especially Dennett, *Elbow Room* and *Consciousness Explained*.

35. See A. Clark, "That Special Something: Dennett on the Making of Minds and Selves," forthcoming in *Dennett Beyond Philosophy*, ed. A. Brook and D. Ross (Cambridge: Cambridge University Press). For some of Dennett's original treatments see Dennett, *Consciousness Explained* and *Kinds of Minds*.

36. In making my case, I have, however, helped myself to a distinction that Dennett would treat with great caution: the distinction between the contents of my current *conscious* awareness and the manifold other goings-on inside my brain (and perhaps, at times, elsewhere). Dennett repeatedly rejects any such clean and crisp separation. He rejects "the assumption that consciousness is a special all-or-nothing property that sunders the universe into two vastly different categories," and adds "we cannot draw the line separating our conscious mental states from our unconscious mental states" (both quotes from Dennett, *Consciousness Explained*, 447). But just what is Dennett rejecting here? Not, I think, the idea that many neural processes operate at what Dennett *himself* originally dubbed the "subpersonal level." In spinning my narrative web, or reporting my experiences, I simply have no access, except of an indirect, scientific kind, to many facts about, for instance, the specifics of my own low-level visual processing, or of my postural adjustment systems, or of many of the processes involved in creative thought. Dennett's point is that within the somewhat smaller space of goings-on of which I have some degree of explicit awareness—where I could, if asked, offer some kind of judgment or report—there is no neat line between what is really, here-and-now conscious and what is not. Moreover, between *all such* potentially reportable goings-on and all the rest, there is no difference so profound as to resist explanation in terms of the flow of information and availability for control of intentional action and verbal report. If this reading is correct, then my use of a conscious/nonconscious distinction is fully compatible with Dennett's own position. For a short treatment by Dennett, which strongly suggests this interpretation, see Dennett, "The Path Not Taken," in *The Nature of Consciousness*, ed. N. Block,

O. Flanagan, and G. Guzeldere (Cambridge, Mass.: MIT Press, 1997). Thanks to Susan Blackmore for drawing this apparent conflict to my attention.

37. For an especially interesting attempt to dismantle such a picture, and for some fascinating reflections on the relation between Buddhist ideas and the "disappearing self," see S. Blackmore, *The Meme Machine* (Oxford: Oxford University Press, 1999) chaps. 17, 18.

38. See D. Edwards, C. Baum, and N. Morrow-Howell, "Home Environments of Inner City Elderly with Dementia: Do They Facilitate or Inhibit Function?" *Gerontologist* 34:1 (1994): 64. In an ongoing project at Washington University, Baum and her colleagues are using neuro-imaging techniques to identify exactly what *kinds* of neural degeneration respond best to environmental restructuring. In this protracted longitudinal study, the group combines imaging with drug therapy with the study and manipulation of the physical home environments, and analysis of the social environment (family, friends, etc.). The goal is to take the social and environmental factors every bit as seriously as the biological ones and try to track the complex interactions between them.

39. See *The Adapted Mind*, ed. J. Barklow, L. Cosmides, and J. Tooby (New York: Oxford University Press, 1992). This is not to suggest, of course, that all evolutionary psychologists speak with one voice. There are as many shades of EP as there are of socialism, for instance. My brief comments address only what seems to be a central tendency in many of the more popular treatments.

40. For a powerful defense of such a view, see P. Griffiths and R. Gray, "Developmental Systems and Evolutionary Explanation," *Journal of Philosophy* 91:6 (1994): 277–305. Unlike Griffiths and Gray, however, I am not yet convinced that the genes do not play a *distinctive* role in this complex matrix. My belief is simply that the genes are, nevertheless, just one element in a kind of culturally modulated cascade, and that our cognitive natures are best seen as products of this much more complex cascade. For some thoughts on the distinctive role of the genes, see A. Clark and M. Wheeler, "Genic Representation: Reconciling Content and Causal Complexity," *British Journal for the Philosophy of Science* 50:1 (1999): 103–35.

Chapter 6

1. See, for example, M. Davies, S. Hawkins, and H. Jones, "Mucus Production and Physiological Energetics in *Patella Vulgata*," *Journal of Mollusc Studies* 56 (1990): 499–503.

2. Special Report, *New Scientist* (March 11, 2000): 26–45.

3. For this example and a general introduction to the notion of swarm intelligence and its technological applications, see E. Bonabeau and G. Théraulaz, "Swarm Smarts," *Scientific American* 282:3 (March 2000): 72–79.

4. Michael Brooks, "Global Brain," *New Scientist* (June 24, 2000).

5. E. Bonabeau and G. Théraulaz, "Swarm Smarts," *Scientific American* 282:3 (March 2000): 76–77. Also, www.iridia.ulb.as.ce/dorigo/ACO/ACO.html (site devoted to ant-based optimization techniques).

6. Michael Brooks, "Global Brain," *New Scientist* (June 24, 2000).

7. The search engine Google uses a related but slightly less intensive procedure. For a comparison, see S. Chakrabarti, B. Dom, D. Gibson, J. Kleinberg, S. R. Kumar, P. Raghaven, S. Rajagopalan, and A. Tomkins, "Hypersearching the Web," *Scientific American*, June 1999.

8. J. Kleinberg, "Authoritative Sources in a Hyperlinked Environment," *Proceedings of the 9th ACM-SIAM Symposium on Discrete Algorithms* (1998). An extended version is in *Journal of the ACM* 46 (1999). It also appears as *IBM Research Report RJ 10076*, May 1997. A pdf version is available on Kleinberg's homepage. References are to the page numbers of the 1997 research report. The quoted passage is from page 2.

9. One page in the root set R is allowed to bring only some set number of new pages into view, and links between pages with the same domain name ("the first level in the URL string associated with a page") are ignored, since these typically serve merely "navigational" functions within one larger document. The only links that are of interest are thus what Kleinberg calls "transverse links": links between pages with different domain names. These and a few additional heuristics are described in Kleinberg, "Authoritative Sources in a Hyperlinked Environment," 6–7.

10. Ibid., 11–12.

11. *New Scientist* 2276 (February, 2001): 17.

12. E. Thelen and L. Smith, *A Dynamic Systems Approach to the Development of Cognition and Action* (Cambridge, Mass.: MIT Press, 1994), 60, 311.

13. For lots more on this, see A. Clark, *Being There: Putting Brain, Body and World Together* (Cambridge, Mass.: MIT Press, 1997).

14. Luis Mateus Rocha, "Adaptive Recommendation and Open-Ended Symbiosis," *International Journal of Human-Computer Studies*. Also available at www.c3.lanl.gov/rocha/

15. James O'Donnell, *Avatars of the Word* (Cambridge, Mass.: Harvard University Press, 1998), 61.

16. For example, my editor at Oxford University Press, Kirk Jensen, points out that under such conditions funding for many journals would be threatened, and

this would have the undesired consequence of making publishing into an underfunded amateur activity. For an opposing viewpoint, see O'Donnell, *Avatars of the Word*.

17. See also Susan Blackmore, *The Meme Machine* (Oxford: Oxford University Press, 1999).

18. Kevin Kelly, *Out of Control* (Reading, Mass.: Perseus Books, 1994), 176.

19. M. Resnick, *Turtles, Termites and Traffic Jams* (Cambridge, Mass.: MIT Press, 1994).

20. For a compelling meditation on these themes, see Steven Johnson, *Emergence: The Connected Lines of Ants, Brains, Cities and Software* (London: Penguin, 2001). Also Kevin Kelly's classic, *Out of Control*.

21. Neil Gershenfeld, *When Things Start to Think* (London: Hodder and Stoughton, 1999), 18.

22. Ibid., 18–31.

23. "El moviento Linux cumple diez años animado por la acogida de la industria," *El Pais (Ciberpais)* 181 (August 23, 2001): 1.

24. Amy Harmon, "For Sale: Free Software," *New York Times*, September 28, 1998, sec. C1, 1–2.

25. Gershenfeld, *When Things Start to Think*, 235.

26. Ibid., 240–41.

27. David Prosser, "Black Box Could Lower Premiums," *Daily Express*, February 20, 2002, 33.

Chapter 7

1. *El Pais (Ciberpais)*, 20 de Julio, 2000, 2.

2. This figure is given by N. Katherine Hayles in *How We Became Post-Human* (Chicago: University of Chicago Press, 1999), 20.

3. Neil Gershenfeld, *When Things Start to Think* (London: Hodder and Stoughton, 1999), 242.

4. Ibid., 243.

5. For full details, see the MIT OpenCourseWare factsheet, published on the web by the MIT News Office. A Google search for MIT OpenCourseWare will hit it first time.

6. Acting together with the British Medical Association and the Soros Foundation. See "Read All About It," *New Scientist* 2299 (July 14, 2001): 5.

7. This and several of the following stories are drawn from a chilling article by Jeffrey Rosen, "The Eroded Self," *New York Times Magazine*, April 30, 2000.

8. Word 97 and Powerpoint 97 are cited by Jeffrey Rosen (see note 7 above).

9. See Bradley J. Rhodes, Nelson Minar, and Josh Weaver, "Wearable Computing Meets Ubiquitous Computing: Reaping the Best of Both Worlds," *Proceedings of the International Symposium on Wearable Computers (ISWC '99)* (October 1999); http://www.media.mit.edu/ rhodes/Papers/wearhive.html; http://citeseer.nj.nec.com/rhodes99wearable.html.

10. Said in a news conference concerning the release of Jini, a new interactive technology hailed as part of the fully networked home in which consumer appliances will communicate with each other and with outside networks (as per the vision of Ubiquitous Computing).

11. Bradley Rhodes et al., "Wearable Computing Meets Ubiquitous Computing."

12. For a lovely account, see Kevin Kelly, *Out of Control* (Reading, Mass.: Perseus Books, 1994), chap. 12.

13. Gershenfeld, *When Things Start to Think*, 94.

14. Rosen, "The Eroded Self."

15. Ibid. Other security measures might include the use of Atguard, a program that watches your online activity, alerts you to monitors and open doors, and sends cookies packing.

16. J. G. Ballard, *Super-Cannes* (London: HarperCollins, 2001), p. 95.

17. For an extended meditation on this theme, see Kelly, *Out of Control*.

18. Gershenfeld, *When Things Start to Think*, 121–22.

19. Both quotes drawn from Donald Norman's discussion titled "No Moments of Silence," in D. Norman, *The Invisible Computer* (Cambridge, Mass.: MIT Press, 1999), 129.

20. For example, an ad for a men's fashion magazine consisting of a black-and-white commercial bar code warped open in the center so as to resemble a vagina and accompanied by the worrying slogan "What every man wants." This image might be laid alongside the bar-coded breasts displayed in chapter 1, as a reminder of what we *don't* want to win a place in (what Donna Haraway called) "man's Family Album."

21. John Pickering, "Human Identity in the Age of Software Agents," in *Cognitive Technology: Instruments of Mind: Proceedings of the 4th International Conference on Cognitive Technology*, ed. M. Beynon, C. Nehaniv, and K. Duatenhahn (Berlin: Springer, 2001), 450.

22. Ibid., 446.

23. Hayles, *How We Became Post-Human*, 115.

24. Pickering, "Human Identity in the Age of Software Agents," 445.

25. See report by Eugenie Samuel, "Gimme, It's Mine," *New Scientist* 2301 (July 28, 2001): 23.

26. Both quotes from *Netfuture* 124 (October 30, 2001).

27. N. Sheppard Jr., "Trashing the Information Highway: White Supremacy Goes Hi-tech," *Emerge* (July–August 1996): 34–40. For some further discussion of related themes, see Thomas Foster, "Trapped by the Body? Telepresence Technologies and Transgendered Performances in Feminist and Lesbian Rewritings of Cyberpunk Fiction," in *The Cybercultures Reader*, ed. D. Bell and B. Kennedy (London: Routledge, 2000), 439–59.

28. I have chosen my words carefully here. Where I might have written of "men posing as women," I have written instead "(biological) men presenting as women." This is because I think (and shall argue) that it is often unclear which of the many identities and aspects available to an individual should be privileged as his/her/its "true self." Better, perhaps, to see the self as the shifting sum of multiple personas adapted to different contexts, constraints, and expectations.

29. It would take me too far afield, and too deep into the philosophy of personhood, to pursue this much further here. But the interested reader might look at some recent treatments such as D. Dennett and N. Humphrey, "Speaking for Our Selves" in D. Dennett, *Brainchildren* (Cambridge, Mass.: MIT Press, 1998), 31–55; Carol Rovane, *The Bounds of Agency* (Princeton, N. J.: Princeton University Press, 1998); Gareth Branwyn, "Compu-Sex: Erotica for Cybernauts," in *The Cybercultures Reader*, 396–402.

30. The phrase, and the account of the Carnegie-Mellon counterattack, is drawn from "Robots Help Humans Defeat Robots," which was a short news piece in the "In Brief" section of *Trends in Cognitive Sciences* 5:12 (2001): 512. The articles for that section were written by Heidi Johansen-Berg and Mark Wrexler.

31. A. Turing, "Computing Machinery and Intelligence," *Mind* 59 (1950): 423–60; reprinted in *The Philosophy of Artificial Intelligence*, ed. M. Boden (Oxford: Oxford University Press, 1990), 40–66. The original Turing Test was named after Alan Turing, who believed that a sufficient degree of behavioral success should be allowed to establish that a candidate system, be it a human or a machine, is a genuine thinker. Turing proposed a test (today known as the Turing Test) that involved a human interrogator trying to spot—from verbal responses—whether a hidden conversant was a human or a machine. Any system capable of fooling the interrogator, Turing proposed, should be counted as a genuinely intelligent agent. Sustained, top-level verbal behavior, if Turing is right, is a sufficient test for the presence of real intelligence.

32. See L. von Ahn, M. Blum, and J. Langford, "Telling Humans and Computers

Apart (Automatically)," Carnegie-Mellon University research paper CMU-CS-02-117. See also http://www.captcha.net.

33. This account of the message's routing is based on Laura Miller's article "One E-Mail Message Can Change the World," *New York Times Magazine*, December 9, 2001.

34. The Slashdot story is based on Steven Johnson's excellent study, *Emergence: The Connected Lives of Brains, Cities and Software* (London: Allen Lane, Penguin Press, 2001), 152–62.

35. Ibid., 153–54.

36. Reported on BBC News, November 24, 2001.

37. FurryMUCK is at www.furnation.com/fgc/. Furry newsgroups include alt.fan.furry, alt.sex.furry, and alt.sex.plushie. For a fairly detailed account, see Julene Snyder's article "Animal Magnetism," which appeared in *Life*, August 26, 1998 (and can be found on the web).

38. H. Moravec, *Mind Children: The Future of Robot and Human Intelligence* (Cambridge and New York: Cambridge University Press, 1998), 17.

39. N. K. Hayles, *How We Became Post-Human*, 291.

Conclusions

1. William Burroughs, *Dead City Radio* (recording available on Island Records, 1990).

2. This debate is especially lively in the area of feminist and literature studies. For a wonderful and ultimately positive window on some of these debates, see N. Katherine Hayles, *How We Became Post-Human* (Chicago: University of Chicago Press, 1999).

Index